166
Advances in Polymer Science

Editorial Board:
A. Abe · A.-C. Albertsson · H.-J. Cantow
K. Dušek · S. Edwards · H. Höcker
J. F. Joanny · H.-H. Kausch · S. Kobayashi
K.-S. Lee · I. Manners · M. Möller · O. Nuyken
S. I. Stupp · U. W. Suter · B. Voit · G. Wegner

Springer

*Berlin
Heidelberg
New York
Hong Kong
London
Milan
Paris
Tokyo*

Polyelectrolytes with Defined Molecular Architecture II

Volume Editor: Manfred Schmidt

With contributions by
V. Abetz · M. Ballauff · C. Seidel · H. Dautzenberg
S. Förster · T. Hofmann · C. Holm · K. Huber
J.F. Joanny · K. Kremer · H. Löwen · A.H.E. Müller
M. Müller · R.R. Netz · W. Oppermann
M. Rehahn · P. Reineker · C. Seidel · M. Schmidt
A.F. Thünemann · T.A. Vilgis · N. Volk
D. Vollmer · R.G. Winkler

 Springer

The series presents critical reviews of the present and future trends in polymer and biopolymer science including chemistry, physical chemistry, physics and material science. It is addressed to all scientists at universities and in industry who wish to keep abreast of advances in the topics covered.

As a rule, contributions are specially commissioned. The editors and publishers will, however, always be pleased to receive suggestions and supplementary information. Papers are accepted for "Advances in Polymer Science" in English.

In references Advances in Polymer Science is abbreviated Adv Polym Sci and is cited as a journal.

The electronic content of APS may be found at http://www.SpringerLink.com

ISSN 0065-3195
ISBN 3-540-00556-0
DOI 10.1007/b10951
Springer-Verlag Berlin Heidelberg New York

Library of Congress Catalog Card Number 61642

This work is subject to copyright. All rights are reserved, whether the whole or part of the material is concerned, specifically the rights of translation, reprinting, re-use of illustrations, recitation, broadcasting, reproduction on microfilms or in other ways, and storage in data banks. Duplication of this publication or parts thereof is only permitted under the provisions of the German Copyright Law of September 9, 1965, in its current version, and permission for use must always be obtained from Springer-Verlag. Violations are liable for prosecution under the German Copyright Law.

Springer-Verlag Berlin Heidelberg New York
Springer Verlag is a part of Springer Science+Business Media
http://www.springer.de

© Springer-Verlag Berlin Heidelberg 2004
Printed in Germany

The use of registered names, trademarks, etc. in this publication does not imply, even in the absence of a specific statement, that such names are exempt from the relevant protective laws and regulations and therefore free for general use.

Typesetting: Stürtz AG, 97080 Würzburg
Cover: Design & Production, Heidelberg
Printed on acid-free paper   02/3020/kk – 5 4 3 2 1 0

Volume Editor

Prof. Dr. Manfred Schmidt
Institut für Physikalische Chemie
Universität Mainz
55128 Mainz, Germany

Editorial Board

Prof. Akihiro Abe
Department of Industrial Chemistry
Tokyo Institute of Polytechnics
1583 Iiyama, Atsugi-shi 243-02, Japan
E-mail: aabe@chem.t-kougei.ac.jp

Prof. Ann-Christine Albertsson
Department of Polymer Technology
The Royal Institute of Technology
S-10044 Stockholm, Sweden
E-mail: aila@polymer.kth.se

Prof. Hans-Joachim Cantow
Freiburger Materialforschungszentrum
Stefan Meier-Str. 21
79104 Freiburg i. Br., Germany
E-mail: cantow@fmf.uni-freiburg.de

Prof. Karel Dušek
Institute of Macromolecular Chemistry, Czech
Academy of Sciences of the Czech Republic
Heyrovský Sq. 2
16206 Prague 6, Czech Republic
E-mail: dusek@imc.cas.cz

Prof. Sam Edwards
Department of Physics
Cavendish Laboratory
University of Cambridge
Madingley Road
Cambridge CB3 OHE, UK
E-mail: sfe11@phy.cam.ac.uk

Prof. Hartwig Höcker
Lehrstuhl für Textilchemie
und Makromolekulare Chemie
RWTH Aachen
Veltmanplatz 8
52062 Aachen, Germany
E-mail: hoecker@dwi.rwth-aachen.de

Prof. Jean-François Joanny
Physicochimie Curie
Institut Curie section recherche
26 rue d'Ulm
75248 Paris Cedex 05, France
E-mail: jean-francois.joanny@curie.fr

Prof. Hans-Henning Kausch
c/o IGC I, Lab. of Polyelectrolytes
and Biomacromolecules
EPFL-Ecublens
CH-1015 Lausanne, Switzerland
E-mail: kausch.cully@bluewin.ch

Prof. S. Kobayashi
Department of Materials Chemistry
Graduate School of Engineering
Kyoto University
Kyoto 606-8501, Japan
E-mail: kobayasi@mat.polym.kyoto-u.ac.jp

Prof. Prof. Kwang-Sup Lee
Department of Polymer Science & Engineering
Hannam University
133 Ojung-Dong
Teajon 300-791, Korea
E-mail: kslee@mail.hannam.ac.kr

Prof. Ian Manners
Department of Chemistry
University of Toronto
80 St. George St.
M5S 3H6
Ontario, Canada
E-mail: imanners@chem.utoronto.ca

Prof. Dr. Martin Möller
Deutsches Wollforschungsinstitut
an der RWTH Aachen e.V.
Veltmanplatz 8
52062 Aachen
E-mail: moeller@dwi.rwth-aachen.de

Prof. Oskar Nuyken
Lehrstuhl für Makromolekulare Stoffe
TU München
Lichtenbergstr. 4
85747 Garching
E-mail: oskar.nuyken@ch.tum.de

Prof. Samuel I. Stupp
Department of Measurement Materials Science
and Engineering
Northwestern University
2225 North Campus Drive
Evanston, IL 60208-3113, USA
E-mail: s-stupp@nwu.edu

Prof. Ulrich W. Suter
Vice President for Research
ETH Zentrum, HG F 57
CH-8092 Zürich, Switzerland
E-mail: ulrich.suter@sl.ethz.ch

Prof. Brigitte Voit
Institut für Polymerforschung Dresden
Hohe Straße 6
01069 Dresden, Germany

Prof. Gerhard Wegner
Max-Planck-Institut für Polymerforschung
Ackermannweg 10
Postfach 3148
55128 Mainz, Germany
E-mail: wegner@mpip-mainz.mpg.de

Advances in Polymer Science
Available Electronically

For all customers with a standing order for Advances in Polymer Science we offer the electronic form via SpringerLink free of charge. Please contact your librarian who can receive a password for free access to the full articles. By registration at:

http://www.SpringerLink.com

If you do not have a standing order you can nevertheless browse through the table of contents of the volumes and the abstracts of each article by choosing Advances in Polymer Science within the Chemistry Online Library.

You will find information about the

– Editorial Board
– Aims and Scope
– Instructions for Authors
– Sample Contribution

at www.springeronline.com using the search function.

Preface

Back in 1996 the German Science Foundation (Deutsche Forschungsgemeinschaft) has launched a nationwide research center on "Polyelectrolytes with defined molecular architecture—synthesis, function and theoretical description" (DFG-Schwerpunkt-Programm 1009: Polyelektrolyte mit definierter Molekülarchitektur—Synthese, Funktion und theoretische Beschreibung). On average 25 research groups from all over Germany and one French group were funded for a total of six years in order to attack and solve long standing problems in the field, to explore new ideas and to create new challenges.

The scientific achievements of this center of research are summarized in the present volumes of Advances in Polymer Science, volume 165 and 166. Financially supported by a "Coordination Funds" the interdisciplinary cooperation between the very many participating research groups was greatly enhanced and has consequently led to contributions involving an unusually large number of authors.

We hope that the center has brought German Polyelectrolyte Research into an international leading position and that it will constitute the nucleus for future activities in this field.

On behalf of all of my colleagues I wish to thank the "Deutsche Forschungsgemeinschaft" for financial and in particular Dr. K.-H. Schmidt and Dr. F.-D. Kuchta for administrative support and to the voluntary reviewers of the proposals, Prof. Blumen, Univ. Freiburg, Prof. Fuhrmann, Univ. Clausthal, Prof. Heitz, Univ. Marburg, Prof. Maret, Univ. Konstanz, Prof. Möller, Univ. Ulm, Dr. Winkler, BASF Ludwigshafen, Prof. Wulf, Univ. Düsseldorf, for their invaluable judgment and advice.

Mainz, February 2003   Manfred Schmidt

Contents

Stiff-Chain Polyelectrolytes
C. Holm, M. Rehahn, W. Oppermann, M. Ballauff   1

Conformation and Phase Diagrams of Flexible Polyelectrolytes
N. Volk, D. Vollmer, M. Schmidt, W. Oppermann, K. Huber 29

Polyelectrolyte Theory
C. Holm, T. Hofmann, J.F. Joanny, K. Kremer, R.R. Netz, P. Reineker,
C. Seidel, T.A. Vilgis, R.G. Winkler 67

Polyelectrolyte Complexes
A.F. Thünemann, M. Müller, H. Dautzenberg, J.F. Joanny, H. Löwen 113

Polyelectrolyte Block Copolymer Micelles
S. Förster, V. Abetz, A.H.E. Müller 173

Author Index Volumes 101–166 211

Subject Index 265

Contents of Volume 165
Polyelectrolytes with Defined Molecular Architecture I

New Polyelectrolyte Architectures
J. Bohrisch, C.D. Eisenbach, W. Jaeger, H. Mori, A.H.E. Müller,
M. Rehahn, C. Schaller, S. Traser, P. Wittmeyer   1

Poly (Vinylformamide-co-Vinylanic)/Inorganic Oxide Hybrid Materials
S. Spange, T. Meyer, I. Voigt, M. Eschner, K. Estel, D. Pleul, F. Simon . . . 43

Polyelectrolyte Brushes
J. Rühe, M. Ballauff, M. Biesalski, P. Dziezok, F. Gröhn,
D. Johannsmann, N. Houbenov, N. Hugenberg, R. Konradi,
S. Minko, M. Motornov, R.R. Netz, M. Schmidt, C. Seidel,
M. Stamm, T. Stephan, D. Usov, H. Zhang 79

Lipid and Polyampholyte Monolayers to Study Polyelectrolyte Interactions and Structure at Interfaces
H. Möhwald, H. Menzel, C.A. Helm, M. Stamm 151

Polyelectrolyte Membranes
R.v. Klitzing, B. Tieke 177

Characterization of Synthetic Polyelectrolytes by Capillary Electrophoretic Methods
H. Engelhardt, M. Martin 211

Author Index Volumes 101–165 249

Subject Index 265

Stiff-Chain Polyelectrolytes

C. Holm[1] · M. Rehahn[2] · W. Oppermann[3] · M. Ballauff[4]

[1] Max-Planck-Institute for Polymer Research, Ackermannweg 10, 55128 Mainz, Germany
[2] Ernst-Berl-Institut für Technische und Makromolekulare Chemie,
Technische Universität Darmstadt, Petersenstrasse 22, 64287 Darmstadt, Germany
[3] Institut für Physikalische Chemie, Technische Universität Clausthal,
Arnold-Sommerfeld-Str. 4, 38678 Clausthal-Zellerfeld, Germany
[4] Universität Bayreuth, Fakultät 2 Biologie/Chemie und Geowissenschaften, LS,
Physikalische Chemie I, Universitätsstraße 30, 95447 Bayreuth, Germany
E-mail: matthias.ballauff@uni-bayreuth.de

Abstract Rod-like polyelectrolytes represent ideal model systems for a comprehensive comparison of theory and experiment because their conformation is independent of the ionic strength in the system. Hence, the correlation of the counterions to the highly charged macroion can be studied without the interference of conformational effects. In this chapter the synthesis and the solution behavior of rigid, rod-like cationic polyelectrolytes having poly(p-phenylene) (PPP) backbones is reviewed. These polymers can be characterized precisely and possess degrees of polymerization of up to $P_n \approx 70$. The analysis of the uncharged precursor polymer demonstrated that the PPP backbone has a high persistence length (ca. 22 nm) and hence may be regarded in an excellent approximation as rod-like macromolecules. The solution properties of the PPP-polyelectrolytes were analyzed using electric birefringence, small-angle X-ray scattering (SAXS) and osmometry. Measurements of the electric birefringence demonstrate that these systems form molecularly disperse systems in aqueous solution. The dependence of electric birefringence on the concentration of added salt indicates that an increase of ionic strength leads to stronger binding of counterions to the polyion. Data obtained from osmometry and small-angle X-ray scattering can directly be compared to the prediction of the Poisson-Boltzmann theory and simulations of the restricted primitive model. Semi-quantitative agreement is achieved.

Keywords Rod-like polyelectrolytes · Poisson-Boltzmann theory · Osmotic coefficient · Electric birefringence · SAXS

1	Introduction...........................	2
2	Theory...............................	5
2.1	Poisson-Boltzmann Theory for the Cylindrical Cell Model.....	5
2.2	Beyond PB and the Cell Model...................	7
3	Synthesis............................	9
4	Solution Properties......................	10
4.1	Electric Birefringence.....................	10
4.2	Osmotic Coefficient.....................	16
4.2.1	Theory and Simulation....................	16
4.2.2	Comparison to Experimental Data...............	18
4.2.3	Comparison to Data taken from Flexible Polyelectrolytes......	20

© Springer-Verlag Berlin Heidelberg 2004

4.3 SAXS, ASAXS 21

5 Conclusion 25

References 25

Abbreviations and Symbols

Latin characters:

a	radius of macroion
f	scattering factor (SAXS)
f'	real part of scattering factor f
f''	imaginary part of scattering factor f
I(q)	scattering intensity of solution
$I_0(q)$	scattering intensity of isolated molecule
L	length of rod-like molecule
n(r)	radial distribution of counterions around rod
q	magnitude of scattering vector
R_M	Manning radius (Eq. (5))
R_0	cell radius
RPM	restricted primitive model
S(q)	structure factor describing interaction between solute molecules

Greek characters:

α	cosin of angle between long axis of rod-like molecule and scattering vector q
β	integration constant of cell model (Eq. (4))
ϕ	osmotic coefficient
ϕ_∞	osmotic coefficient in Manning limit
κ	screening constant
λ_B	Bjerrum length (Eq. (2))
Π	osmotic pressure
ξ	charge parameter (Eq. (1))

1
Introduction

The understanding of flexible polyelectrolytes in dilute solutions of low ionic strength still presents a considerable challenge in macromolecular science despite of many decades of research [1–5]. This is due to the long-range nature of the Coulombic forces between the charged macromolecules. In the case of flexible polyelectrolytes, a decrease of the ionic strength may lead to an expansion of the coils due to strong intramolecular forces as well as to stronger intermolecular electrostatic interactions. Conformationally rigid, rod-like polyelectrolytes, on the other hand, remain in their extended chain conformation regardless of the ionic strength of the system. Because conformative effects are ruled out here, only the Coulombic interactions determine the solution properties of these polymers.

Based on these considerations, a number of studies have been performed using naturally occurring rod-like helical polyelectrolyte systems such as DNA or xanthane [6, 7]. However, at very low ionic strengths and at elevated temperatures, the helical conformation and thus the rod-like shape is lost. Moreover, systematic variation of the charge density, i.e., the number of ionic groups per unit length, is not possible using these biopolymers. Therefore, the development of well-defined synthetic rod-like polyelectrolytes is necessary to analyze quantitatively intermolecular interactions, the correlation of the counterions with the macroion, and structure formation in solution depending on ionic strength, temperature, and polyelectrolyte concentration.

The first syntheses of rod-like polyelectrolytes were published in the early eighties [8, 9]. They were based on poly(1,4-phenylenebenzobisoxazoles) and poly(1,4-phenylenebenzobisthiazoles). In the recent decade, we [10–14] and others [15,16] developed various efficient precursor routes to synthetic rod-like poly(p-phenylene) (PPP) polyelectrolytes which take advantage of both the concept of solubilizing side chains [17,18] and the efficient Pd-catalyzed aryl-aryl coupling reaction [19–21]. This progress was rendered possible mainly by (*i*) the high tolerance of the Pd-catalyzed polycondensation reactions toward functional groups in the starting materials and (*ii*) the outstanding thermal and chemical stability of the PPP backbone which allows transformation of the uncharged precursor PPPs into polyelectrolytes by a variety of organic reactions. By these precursor routes, high-molecular-weight carboxylated and sulfonated PPP polyelectrolytes with homogeneous constitution and known degrees of polymerization (P_n) have been prepared first. The polymers thus available combine exceptional hydrolytic, thermal and chemical stability with a high charge density of up to four ionic groups per *p*-phenylene repeating unit. Hence, these systems present nearly ideal model polyelectrolytes to be studied in solution.

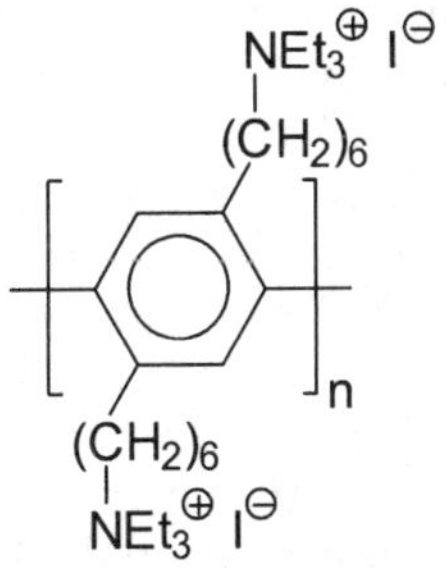

Scheme 1 Chemical structure of polyelectrolyte **PPP-1**

Recently, the stiff-chain polyelectrolytes termed **PPP-1** (Scheme1) and **PPP-2** (Scheme2) have been the subject of a number of investigations that are reviewed in this chapter. The central question to be discussed here is the correlation of the counterions with the highly charged macroion. These correlations can be detected directly by experiments that probe the activity of the counterions and their spatial distribution around the macroion. Due to the cylindrical symmetry and the well-defined conformation these polyelectrolytes present the most simple system for which the correlation of the counterions to the macroion can be treated by analytical approaches. As a consequence, a comparison of theoretical predictions with experimental results obtained in solution will provide a stringent test of our current model of polyelectrolytes. Moreover, the results obtained on **PPP-1** and **PPP-2** allow a refined discussion of the concept of "counterion condensation" introduced more than thirty years ago by Manning and Oosawa [22, 23]. In particular, we can compare the predictions of the Poisson-Boltzmann mean-field theory applied to the cylindrical cell model and the results of Molecular dynamics (MD) simulations of the cell model obtained within the restricted primitive model (RPM) of electrolytes very accurately with experimental data. This allows an estimate when and in which frame this simple theory is applicable, and in which directions the theory needs to be improved.

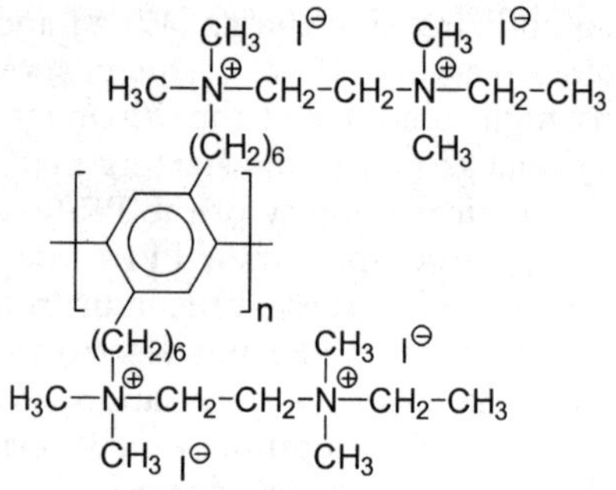

Scheme 2 Chemical structure of polyelectrolyte **PPP-2**

The review is organized as follows: While Sect. 2 gives an overview over the relevant polyelectrolyte theory, Sect. 3 describes the synthetic routes that lead to the polyelectrolytes **PPP-1** and **PPP-2**. Then, in Sect. 4, the solution properties of **PPP-1** and **PPP-2** are discussed. Here the electric birefringence showed for the first time that the polyelectrolyte **PPP-1** forms molecularly disperse solutions. Moreover, transport properties in general can be compared to the results obtained by birefringence. The osmotic coefficient as well as small-angle X-ray scattering have been chosen as further experimental observables to be discussed because they give the most conclusive insight into the distribution of the counterions around the cylindrical macroion. Our conclusion will summarize the results obtained so far on stiff-chain polyelectrolytes and we briefly mention the direction of further research in Sect. 5.

2
Theory

In general, one of the characteristics of rod-like polyelectrolytes is the charge (Manning) parameter ξ which for monovalent counterions is defined through the ratio of the Bjerrum length λ_B to the contour distance per unit charge b [22–24]:

$$\xi = \frac{\lambda_B}{b} \tag{1}$$

with λ_B being defined through

$$\lambda_B = \frac{e^2}{4\pi\varepsilon_0 \varepsilon k_B T} \tag{2}$$

where e is the unit charge, ε the dielectric constant of the medium and ε_0, k_B and T have their usual meanings. In the following we only consider strongly charged polyelectrolytes with $\xi > 1$. To keep the treatment as simple as possible, we mostly consider salt-free solutions. Moreover, we consider the macroion to be infinitely stiff, i.e., all effects due to flexibility or curvature of the macroion are not taken into account.

2.1
Poisson-Boltzmann Theory for the Cylindrical Cell Model

The cell model is a commonly used way of reducing the complicated many-body problem of a polyelectrolyte solution to an effective one-particle theory [24–30]. The idea depicted in Fig. 1 is to partition the solution into sub-volumes, each containing only a single macroion together with its counterions. Since each sub-volume is electrically neutral, the electric field will on average vanish on the cell surface. By virtue of this construction different sub-volumes are electrostatically decoupled to a first approximation. Hence, the partition function is factorized and the problem is reduced to a single-particle problem, namely the treatment of one sub-volume, called "cell". Its shape should reflect the symmetry of the polyelectrolyte. Reviews of the basic concepts can be found in [24–26].

For a solution of N rod-like polyelectrolytes with density $n=N/V$ and rod length L this gives a cylindrical cell with the cell radius R_0 being fixed by the condition $\pi R_0^2 LN/V=1$ (for the definition of these quantities, see Fig. 1). The theoretical treatment is much simpler after neglecting end effects at the cylinder caps. This is equivalent to a treatment of rods of infinite length after mapping to the correct density. The analytical description of this model proceeds within the Poisson-Boltzmann (PB) approximation: the ionic degrees of freedom are replaced by a cylindrical density $n(r)$ that describes the radial distribution of counterions around the macroion. The distribution $n(r)$ is locally proportional to the Boltzmann factor [24–30]. In doing so all correla-

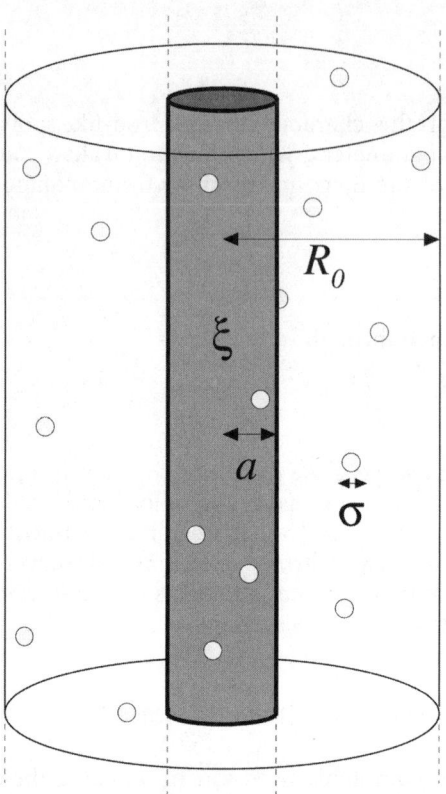

Fig. 1 Scheme of the PB cell model: The rod-like macroion with radius a is confined in a cell of radius R_0 together with its counterions. The charge density of the macroion is characterized by the charge parameter ξ (see Eq. (1)). See Sect. 2.1 for further explanation

tions among the counterions are neglected and the counterions behave as pointlike objects.

The PB-differential equation may be solved analytically for salt-free solution [27, 28], and $n(r)$ is given by [28]

$$\frac{n(r)}{n(R_o)} = 2\left\{\frac{\beta}{\kappa r \cos[\beta \ln(r/R_M)]}\right\}^2 \tag{3}$$

From the known parameters ξ, R_o, and the radius of the macroion a, the first integration constant β has to be obtained through a numerical solution of the transcendental equation

$$\arctan\left(\frac{\xi-1}{\beta}\right) + \arctan\left(\frac{1}{\beta}\right) - \beta \ln\left(\frac{R_o}{a}\right) = 0 \tag{4}$$

The second integration constant R_M may be regarded as the radial distance within the counterions are condensed [28, 29]. It follows as

$$R_M = a\exp\left[\frac{1}{\beta}\arctan\left(\frac{\xi-1}{\beta}\right)\right] \quad (5)$$

The screening constant κ and the number density $n(R_o)$ of counterions at the cell boundary are related through $\kappa=8\pi\lambda_B n(R_o)=4(1+\beta^2)/R_o^2$.

The osmotic coefficient ϕ is defined as

$$\phi = \frac{\Pi}{\Pi_{id}} \quad (6)$$

where Π_{id} is the ideal osmotic pressure of the counterions, and the value of the osmotic pressure Π can be conveniently calculated from the counterion density at the cell boundary, $\Pi=n(R_o)$ kT, since the electric field value vanishes there due to electroneutrality [30]. The osmotic coefficient for systems with $\xi>1$ and monovalent counterions follows directly as

$$\phi = \frac{1+\beta^2}{2\xi} \quad (7)$$

In the limit of infinite dilution, $R_0\to\infty$, one finds $\beta\to 0$, and PB theory recovers the well known Manning [22] limiting law $\phi_\infty = \frac{1}{2\xi}$. At finite densities, however, ϕ is always larger. Experimental measurements of the osmotic pressure in salt-free solutions can hence directly be compared to the predictions of Eq. (7). In addition, the distribution function $n(r)$ given by Eq. (3) can be used to calculate the scattering intensity in small-angle X-ray and neutron scattering (cf. Sect. 4.3). A comparison of theory and experiment therefore provides a stringent test of the underlying assumptions of the PB-theory within the cell model. In addition simulations of the cell model can go beyond the mean-field solution and solve the cell model within the restricted primitive model (RPM) of electrolytes. In this model the ions have a finite diameter σ and the full Coulomb interactions are taken into account. Deviations will therefore give important hints at which points the theoretical treatment needs to be improved.

2.2
Beyond PB and the Cell Model

As already stated in the preceding section, the PB equation neglects ion size effects and interparticle correlations. One route to improve the theory can be done on a density functional level. The PB equation can be derived via a variational principle out of a local density functional [25, 31]. This is also a convenient formulation to overcome its major deficiencies, namely the neglect of ion size effects and interparticle correlations. The Hohenberg-Kohn theorem gives an existence proof of a density functional that will produce the correct density profile upon variation. However, it does not specify its form. Various ways of incorporating ion size effects [32, 33] and correlation effects [31, 34, 35] have been suggested, local and non-local ones. Recently, we were able to derive a stable local ion correlation correction term on the basis of the Debye-Hückel-Hole-Cavity (DHHC) theory [31] which compares very favorable to simulations performed on the rod-like cell model, including multivalent counterions. The correlations generally produce a larger density of the ions near the rod, and lead thus to a lower osmotic coefficient than the PB theory predicts.

Integral equations theories are another approach to incorporate higher order correlations, and consequently also lead to lowered osmotic coefficients. There are numerous variants of these theories around which differ in their used closure relations and accuracy of the treatment of correlations [36]. They work normally very well at high electrostatic coupling and high densities, and are able to account for overcharging, which was first predicted by Lozada-Cassou et al. [36] and also describe excluded volume effects very well, see Refs. [37] for recent comparisons to MD simulations.

Another attempt to go beyond the cell model proceeds with the Debye-Hückel-Bjerrum theory [38]. The linearized PB equation is used as a starting point, however ion association is inserted by hand to correct for the non-linear couplings. This approach incorporates rod-rod interactions and should thus account for full solution properties. For the case of added salt the theory predicts an osmotic coefficient below the Manning limiting value, which is much too low. The same is true for a simplified version of the salt free case.

Recently, Nyquist et al. [39] tried to develop a theory for rods of finite size. These authors used a two-state model for the counterions and employed a random phase approximation in order to calculate the osmotic coefficient ϕ of rod-like polyelectrolytes [39]. An important goal of this work was to reproduce in the zero density limit the correct osmotic coefficient of 1 instead of the Manning limiting value which is due to the unphysical infinite rod assumption employed. The model presented in ref. [39], however, seems to overestimate considerably the osmotic coefficient when compared to experimental data (see below Sect. 4.2).

Another viable method to compare experiments and theories are simulations of either the cell model with one or more infinite rods present or to take a solution of finite semi-flexible polyelectrolytes. These will of course capture all correlations and ionic finite size effects on the basis of the RPM, and are therefore a good method to check how far simple potentials will suffice to reproduce experimental results. In Sect. 4.2, we shall in particular compare simulations and results obtained with the DHHC local density functional theory to osmotic pressure data. This comparison will demonstrate to what extent the PB cell model, and furthermore the whole coarse grained RPM approach can be expected to hold, and on which level one starts to see solvation effects and other molecular details present under experimental conditions.

3
Synthesis

A comprehensive overview on the different synthetic strategies leading to PPP polyelectrolytes is given in Chapter 3, p. 67 ff. Therefore a short description will be suffice here in order to demonstrate how the polymers **PPP-1** and **PPP-2** have been prepared which are under particular consideration here. Prior to this, however, a short comment may be indicated why we selected PPP as the parent system for our stiff-chain polyelectrolytes.

The PPP is an intrinsically rod-like molecule, i.e. it does not need a potentially labile helical superstructure to assume its rod-like shape. Moreover, well-defined PPPs are readily available by combining the Pd-catalyzed Suzuki coupling with the concept of solubilizing side chains. The PPPs are perfectly inert against hydrolysis and other processes possible in aqueous media. Finally, the PPPs are also inert against many other chemical reactions that might be useful to generate functional groups in its lateral side groups.

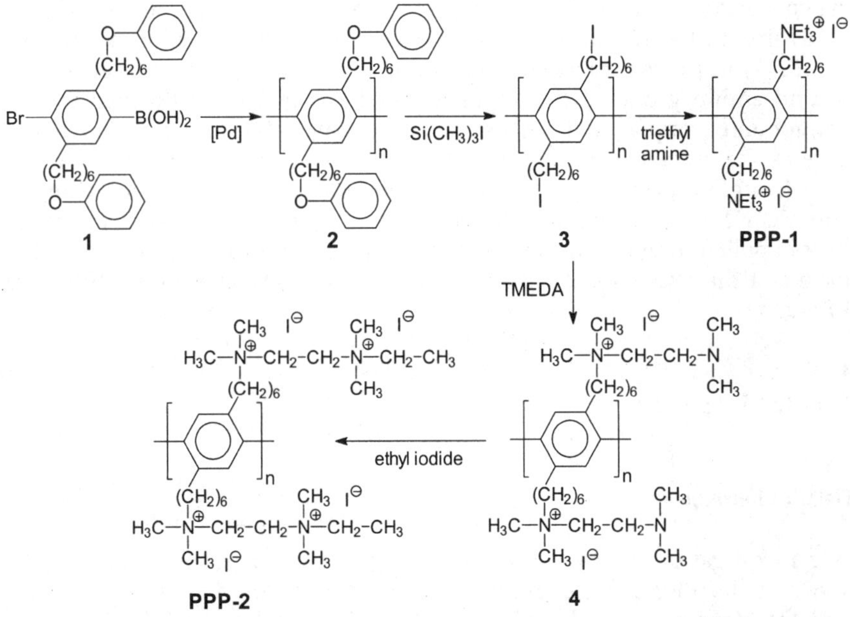

Scheme 3 Synthesis of the PPP polyelectrolytes **PPP-1** and **PPP-2** [10–12]

This latter aspect is of special importance here because it is difficult and less secure to determine molar masses or molar mass distributions of polyelectrolytes. The molecular weight and the contour length of the PPP are needed, however, for a profound interpretation of the observed solution properties. The PPP-based systems open up the opportunity to realize a precursor strategy: the Pd-catalyzed polycondensation process can produce a non-ionic PPP derivative first. This intermediate can readily be characterized using all methods usually applied for polymer analysis. Due to the inertness of the PPP backbone these precursor polymers can be converted via efficient and selective macromolecular substitution processes into well-defined polyelectrolytes. This step can be done without loosing the information on molar mass. For **PPP-1** and **PPP-2**, this concept was realized as shown in Scheme 3 [11–13].

2,5-Bis(ω-phenoxyhexyl)-4-bromobenzene boronic acid **1** was the required AB type monomer to prepare **PPP-1** and **PPP-2** via a precursor route. When carefully purified, its Pd-catalyzed polycondensation in the heterogeneous system toluene (or THF) / 1 M aqueous Na_2CO_3, using approx. 0.5 mol-% of $[Pd(PPh_3)_4]$ as the precursor complex of the catalytically active species, leads to a readily soluble, constitutionally homogeneous precursor polymer **2** having values of P_n of up to 70 (as measured by membrane osmometry [11]). The subsequent ether cleavage **2** $\rightarrow$ **3** using $(H_3C)_3Si$-I in CCl_4 proceeds completely, provided strictly water-free conditions are adhered to. Using the "activated" precursor polymer **3**, **PPP-1** is easily available via conversion with triethyl amine [12]. To obtain **PPP-2**, precursor polymer **3** has first to be treated with a large excess of tetramethylethylene diamine (TMEDA) to prevent cross-linking during quaternization. In a final step the second amino group of the TMEDA moieties can be transformed into an ammonium group as well. This step can be done using for example ethyl iodide as the reagent [13]. For all intermediates as well as for the final PPP polyelectrolytes, a full constitutional analysis could be performed using ^1H and ^{13}C NMR spectroscopy. Moreover, due to the careful work-up procedures applied, one can be sure that the molecular information acquired by means of the precursor **2** remains valid also for polyelectrolytes **PPP-1** and **PPP-2**.

4
Solution Properties

4.1
Electric Birefringence

The investigation of the electric birefringence is an excellent tool for the study of the PPP polyelectrolytes because this method is highly sensitive and therefore particularly suited for very dilute solutions [40–42]. At low field strength, the birefringence observed in solutions or suspensions of non-interacting molecules or particles rises with the square of field strength (Kerr's law) and in proportion with concentration [43]:

$$\Delta n = K_{sp} c E^2 \tag{8}$$

The proportionality constant K_{sp} (specific Kerr constant) depends on the optical anisotropy of the molecules and on the anisotropy of their electric

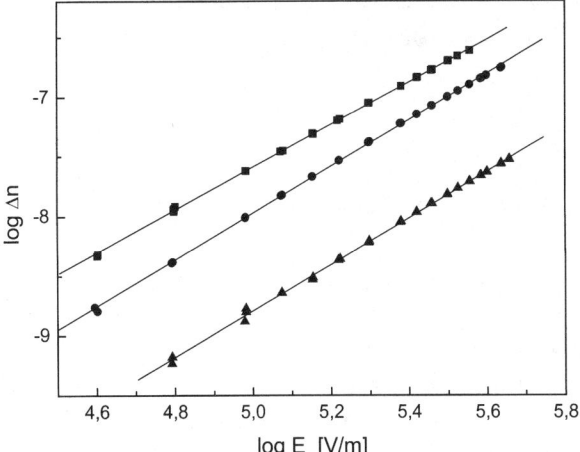

Fig. 2 Dependence of the electric birefringence on field strength for solutions of **PPP-1** having $c=1.15 \cdot 10^{-4}$ bmol/L, squares; $P_n=65$; circles: $P_n=40$; triangles $P_n=20$. All data have been taken from ref. [49]

polarizability. The optical anisotropy of conformationally rigid polymers, in particular PPP polyelectrolytes, is a function of chain length only and not perceptibly affected by external conditions, contrary to flexible polyelectrolytes. The electric polarizability of polyelectrolytes as well as its anisotropy is extremely large, which results in high degrees of orientation in an electric field. This fact can be traced back to an easy displacement of counterions relative to the polyion [44–47]. It therefore comprises the interaction between counterions and the polyion which is a point of considerable interest for a basic understanding of polyelectrolytes in solution.

Three samples of **PPP-1** having number-average degrees of polymerization P_n of 20, 40, and 65 were studied [48, 49]. The P_n had been obtained by membrane osmometry on the uncharged precursors (cf. Sect. 3). Since the polymers were made via palladium-catalyzed polycondensation described in Sect. 3, a Schulz-Flory distribution of molecular weights can be anticipated. The sample having $P_n=65$, however, was obtained by fractionating a sample of lower degree of polymerization and using the high molecular weight fraction. The polydispersity of this particular sample is probably narrower than that of the other two specimens having P_n 20 and 40, respectively.

The dependence of the electric birefringence on field strength is shown in Fig. 2. The monomolar concentration is $1.15 \cdot 10^{-4}$ mol/L (80 mg/L). The data fall on straight lines having slopes close to 2. This means that Kerr's law is valid. An estimate of the saturation value of the birefringence gave a number of more than $2 \cdot 10^{-6}$. This is by at least a factor of 10 larger than the highest values actually observed. At a given field strength, the electric birefringence of the sample with $P_n=65$ is more than an order of magnitude larger than of the one having $P_n=20$. Since the mass concentrations are the same, this must

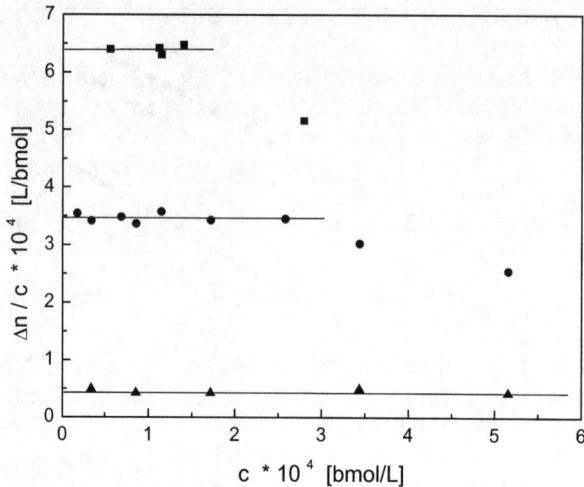

Fig. 3 Concentration dependence of the specific electric birefringence at $E=2\cdot10^5$ V/m, squares: P_n=65; circles: P_n=40; triangles: P_n=20. All data have been taken from ref. [49]

be caused by a much larger orientation of the PPP with higher molar mass, indication of the well-known fact that the ionic polarizability of polyelectrolytes rises markedly with increasing chain length.

The concentration dependence of the electric birefringence was studied at several fixed field strengths. The data obtained at $E=2\cdot10^5$ V/m are shown in Fig. 3, where the specific birefringence, $\Delta n/c$, is plotted versus concentration [49]. At low concentrations, $c\leqslant(2-5)\cdot10^{-4}$ bmol/l or 0.15–0.35 g/l, depending on the degree of polymerization, $\Delta n/c$ is constant; this means that the birefringence is additive with regard to concentration. For the high molecular weight samples, some drop is observed at concentrations exceeding this range.

The proportionality of electric birefringence with concentration (Fig. 3) as well as the clear molar mass dependence (Fig. 3) are important observations since they strongly suggest that the PPP polyelectrolytes studied form molecular solutions without associations or aggregates, and that intermolecular interactions are not of significance under the experimental conditions applied (concentrations below 0.15–0.35 g/l).

For another poly-p-phenylene system it is reported that aggregation to defined cylindrical micelles occurs in aqueous solution [15, 16]. In these systems the ionic groups (sulfonate groups) are directly attached to the phenylene units. Moreover, long n-alkyl side chains are attached to the PPP backbone. The polyelectrolytes **PPP-1** considered here have the trialkyl ammonium groups linked to the backbone via a hexamethylene spacer. It is obvious that the spatial requirement of these substituents prevents the macromolecules from forming such aggregates.

In the following, we shall discuss the influence of low molecular weight electrolytes (salt) on the properties of solutions of the PPP polyelectrolytes.

The quantities considered are the electric birefringence as well as the electric conductivity of the solutions, both being indicative of the mobility of counterions in the system. As pointed out in the theoretical part, the counterions can either be subdivided in condensed or free ions (in the Manning picture), or their binding to the polyion is treated within the Poisson Boltzmann or similar approaches (see Sect. 2). With regard to electric birefringence or, specifically, the polarization mechanism, different opinions exist on whether the displacement of condensed (tightly bound) counterions or the displacement or deformation of the ion cloud (loosely bound ions) are the essential cause of the high anisotropy of polarizability. Early treatments focussed mainly on the condensed ions, considering them moving in a potential trough along the polyion [45–47, 50]. Some refinement was made by allowing for an exchange between bound and unbound counterions [51]. Other theories were based on the polarization of the ion atmosphere [52–54]. Monte Carlo simulations also indicated that the major part of the induced dipole moment results from the displacement of the ion cloud [55].

Electric conductivity gives another measure of ionic mobility, and it is thus worthwhile to compare the changes of electric birefringence and electric conductivity occurring when salts are added to solutions of the PPP polyelectrolytes [49]. In Fig. 4 a, $\Delta n/c$ is plotted versus the concentration of added NaCl, NaI, $N(C_2H_5)_4I$, or Na_2SO_4. The solution contains the PPP polyelectrolyte with $DP_n=40$ at a concentration of $8.6 \cdot 10^{-5}$ bmol/L. There is a distinct decrease of $\Delta n/c$ with rising salt concentration when a certain threshold concentration is exceeded. The data for the three different salts consisting of monovalent ions coincide closely. This rules out any ionspecific effect. However, the decrease occurs at a perceptibly lower salt concentration when Na_2SO_4 is employed. Hence the valency of the counterions is obviously crucial.

Figure 4b shows the influence of added salt on the conductivity contribution of the polyelectrolyte, $\Delta\kappa$. This quantity is obtained by subtracting the conductivity of the pure salt solution (concentration c_{salt}) from that of the solution containing polyelectrolyte and added salt (concentrations c and c_{salt}): $\Delta\kappa=\kappa(PE + salt) - \kappa(salt)$. $\Delta\kappa$ exhibits a similar course as $\Delta n/c$. Again, the curves for NaCl, NaI, and $N(C_2H_5)_4I$ coincide, while that of Na_2SO_4 shows the decrease at a lower salt concentration. It is particularly noteworthy, that the observed changes of $\Delta n/c$ and $\Delta\kappa$ occur in the same concentration range. To elaborate further the similarity of $\Delta\kappa$ and Δn upon addition of the electrolyte, these two quantities were plotted against each other in Fig. 5. The graph contains the same data as those used in Figs. 4a and 4b. A reasonably good correlation is obtained showing that, irrespective of the valency of the counterions, concomitant changes of $\Delta\kappa$ and Δn are observed.

A decrease of Δn or $\Delta\kappa$ upon addition of salt is quite common for flexible polyelectrolytes (see e.g. [56, 57]). It is generally interpreted as being a consequence of the conformational change brought about by the rise of ionic strength. When the coil size is reduced, the optical and electrical polarizability of the polyions is diminished. This leads to the observed drop of electric birefringence. Coiling of the polyion can also lead to an increase of counteri-

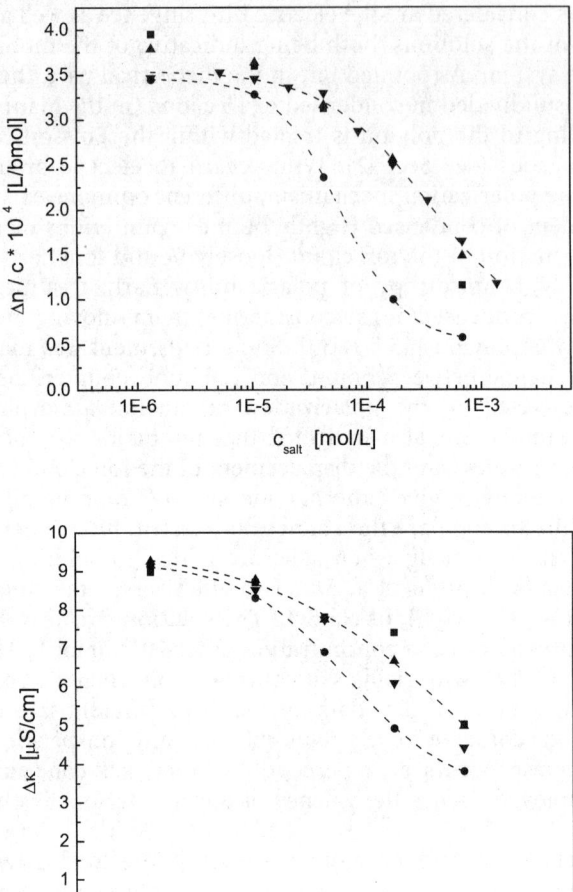

Fig. 4a,b Specific electric birefringence measured at $E=2 \cdot 10^5$ V/m (top) and conductivity contribution of the polyelectrolyte $\Delta\kappa$ (bottom) versus concentration of added salt, squares: NaCl; triangles up: NaI; triangles down: $N(C_2H_5)_4I$; circles: Na_2SO_4. All data have been taken from ref. [49]

on condensation, which would explain the corresponding effect on conductivity. However, this interpretation cannot hold for the systems under consideration here, which are conformationally rigid. Since a change of polyion size and shape is excluded, the phenomena observed must be solely due to changes of polyion-counterion interactions, and the use of stiff-chain polyelectrolytes is particularly advantageous to study these phenomena.

To interpret the decrease of $\Delta\kappa$ it is assumed that a larger fraction of counterions will condense on the polyion, when the ionic strength is in-

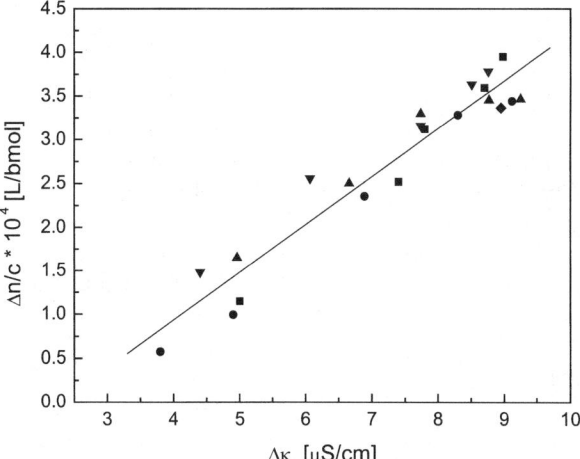

Fig. 5 Correlation between $\Delta n/c$ and $\Delta\kappa$, squares: NaCl; triangles up: NaI; triangles down: $N(C_2H_5)_4I$; circles: Na_2SO_4; diamonds: salt free. All data have been taken from ref. [49]

creased [49]. This reduces the conductivity contribution of the polyelectrolyte. As an equivalent explanation one can conclude that the mobility of ions within the ion cloud must be substantially lower because they are bound more strongly to the polyion. Both interpretations have in common that the conductivity contribution of the ion cloud as a whole is reduced. The fact that Δn decreases in the same manner as $\Delta\kappa$ indicates that the electric polarizability anisotropy is largely determined by the ions in the ion cloud and not by the layer of condensed ions. When divalent counterions are present, as it is the case upon addition of Na_2SO_4, the observed drops of $\Delta\kappa$ and Δn are more pronounced and occur at a lower concentration of the added salt. This is what one would expect from the basic ideas of counterion condensation theory (see Sect. 2).

The experimental results discussed above show that some important information on the origin of the high anisotropy of the ionic polarizability of polyelectrolytes could be deduced from a comparison of the changes of electric birefringence and electric conductivity. As the PPP polyelectrolytes studied have a conformationally rigid backbone, it was possible to perform measurements at different salt concentrations without inducing conformational changes. This is an essential advantage over studying flexible polyelectrolytes, and it is an important prerequisite to arrive at a clear-cut interpretation.

4.2
Osmotic Coefficient

As already indicated in Sect. 2, the osmotic coefficient ϕ provides a sensitive test for the various models describing the electrostatic interaction of the counterions with the rod-like macroion. It is therefore interesting to first compare the PB theory to simulations of the RPM cell model [26, 29] in order to gain a qualitative understanding of the possible failures of the PB theory. In a second step we compare the first experimental values ϕ obtained on polyelectrolyte **PPP-1** [58] quantitatively to PB theory and simulations [59].

4.2.1
Theory and Simulation

Figure 6 gives the comparison of the osmotic coefficient predicted by the PB-theory to simulated data [26, 60]. The simulation system is not strictly a cell system, rather we considered an infinite array of parallel aligned rods which sit on a hexagonal lattice. The rod diameter a was of the same size as the counterions σ, the line charge density λ had the value $\lambda=0.9593\ e_0/\sigma$, and the density and the Bjerrum length was varied. For details of the simulations we refer to Ref. [26, 60].

The first set of simulation has been done for monovalent counterions of size σ and three values of the reduced Bjerrum length $\lambda_B/\sigma=1, 2, 3$. Several findings may be noted: The osmotic coefficient from the simulations is always smaller than the PB prediction but for low density both values converge. This also illustrates that the Manning limiting law becomes asymptot-

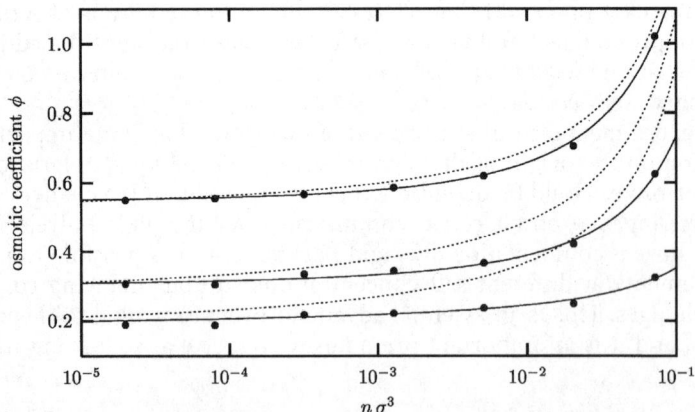

Fig. 6 Osmotic coefficient ϕ versus reduced density n/σ^3 for monovalent counterions. Heavy dots mark the measurements, while the solid lines are fits which merely serve to guide the eye. The dotted lines are the prediction of PB theory. From top to bottom the Bjerrum length λ_B/σ varies as 1,2,3. The errors in the measurement are roughly as big as the dot size [29]

ically correct for dilute systems. Upon increasing the density, the osmotic coefficient rises weaker than the PB prediction. This is more pronounced for systems with higher Bjerrum length, and consequently with a higher Manning parameter. It is due to enhanced counterion condensation which has been reported in [26, 29, 60].

Notice that this has a very remarkable side-effect. Over a considerable range of densities the measured osmotic coefficient is much closer to the limiting law than to the actual PB prediction. This makes the Manning limit look much more accurate than it really is. However, the surprising effect should not be over-interpreted, since the underlying reason is nothing but a fortunate cancellation of two contributions of approximately the same size but opposite sign, which are not contained in the limiting laws.

It is also interesting to investigate complementary systems in which the values of λ_B/σ and valence v have been interchanged to keep the product ξv fixed. The integrated ion distribution function $P(r)$ can be shown to depend only on the product ξv in the PB theory [29]. However, at given density the cell radius depends on the valence and so does β. Therefore the osmotic coefficient does no longer universally depend on this product. The Manning limiting law for multivalent counterions, however, does again only depend on this product, i.e.

$$\phi_\infty = \frac{1}{2v\xi} \qquad \text{(a)}$$

Figure 7 summarizes the results of simulations on the multivalent systems with $v=1, 2, 3$, which yield the same values of ξv as the monovalent ones in-

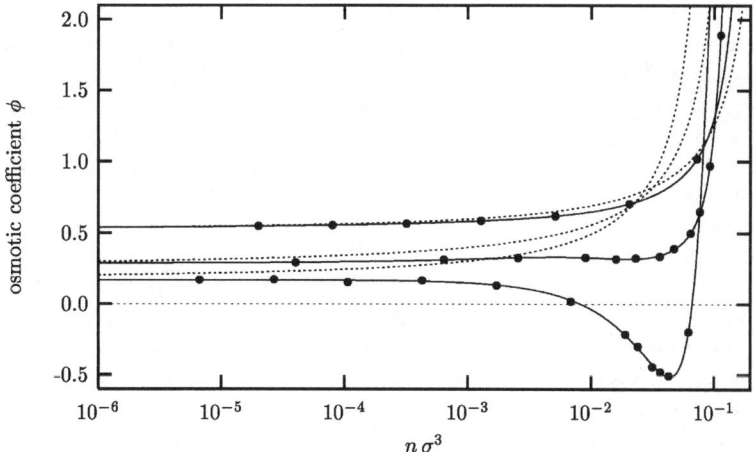

Fig. 7 Osmotic coefficient as a function of reduced density n/σ^3 for different valences. Heavy dots mark the measurements, while the solid lines are fits which merely serve to guide the eye. The dotted lines are the prediction of PB theory. From top to bottom the counter ion valence v varies like 1,2,3, which gives the same value of ξv as the curves in Fig. 6. The errors in the measurement are roughly as big as the dot size [29]

vestigated before in Fig. 6 [29]. The most striking feature is the appearance of a negative osmotic coefficient in a certain density region of the trivalent counterions. If the constraint of a fixed rod-separation were to be replaced, the system would phase separate. This means that attractive interactions must be present between the rods. Similar observations have been reported in Refs. [26, 29, 60, 62–67].

Contrary to the simulations, the osmotic coefficient from the PB theory is always positive. This is clear, because on the one hand we know that in PB theory the pressure is proportional to the density of ions at the cell boundary, which is bounded by zero from below [24], and on the other hand it is the consequence of the rigorous proofs that such attractive interactions are absent on the PB level [61–63]. Finally it should be noted that the above measurements can not be used to infer that attractive forces between charged rods require the counterions to be at least trivalent. The reason is twofold: First, at given valence one can vary Bjerrum length and line charge density. Increasing the Manning parameter will lead to negative pressure in the divalent system. Second, keeping all interaction potentials fixed, the radius a of the charged rod is a relevant observable, as has been demonstrated in Refs. [60, 66]. Hence, a general statement about presence or absence of attractive interactions is difficult, since a five-dimensional parameter space is involved: $(\lambda, \lambda_B, a, v, n)$, where λ denotes the line charge density of the rod.

4.2.2
Comparison to Experimental Data

Up to now, only two sets of data of the osmotic coefficient of rod-like polyelectrolytes in salt-free solution are available: 1) Measurements by Auer and Alexandrowicz [68] on aqueous DNA-solutions, and 2) Measurements of polyelectrolyte **PPP-1** in aqueous solution [58]. A critical comparison of these data with the PB-cell model and the theories delineated in Sect. 2.2 has been given recently [59]. Here it suffices to discuss the main results of this analysis displayed in Fig. 8. It should be noted that the measurements by the electric birefringence discussed in Sect. 4.1 are the most important prerequisite of this analysis. These data have shown that **PPP-1** form a molecularly disperse solution in water and the analysis can therefore assume single rods dispersed in solution [49].

First of all, the comparison of the PB-theory and experiment shown in Fig. 8 proceeds virtually without adjustable parameters. The osmotic coefficient ϕ is solely determined by the charge parameter ξ which in turn is fixed by chemistry, the rod radius a, which has been deducted from SAXS-measurements (see below Sect. 4.3), and the polyelectrolyte concentration. The latter parameter determines the cell radius R_0 (see the discussion in Sect. 2.1) Figure 8 summarizes the results. It shows the osmotic coefficient of an aqueous **PPP-1** solution as a function of counterion concentration as predicted by Poisson-Boltzmann theory, the DHHC correlation-corrected treatment from Sect. 2.2, Molecular Dynamics simulations [29, 59] and experiment [58].

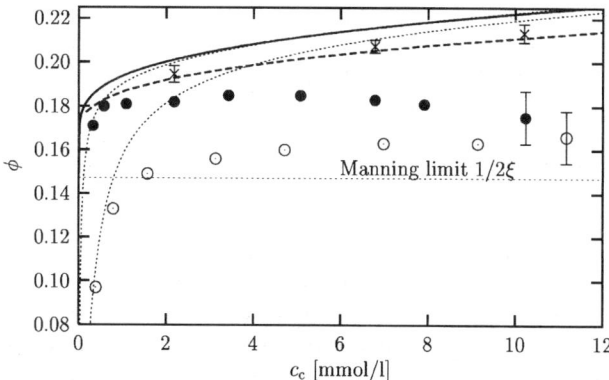

Fig. 8 Osmotic coefficient as a function of counterion concentration c_c for the poly(p-phenylene) systems described in the text. The solid line is the PB prediction of the cylindrical cell-model, the dashed curve is the prediction from the correlation corrected PB theory from Ref. [58]. The full dots are experiments with iodine counterions and the empty dots are results of MD simulations described in ref. [29,59]. The Manning limiting value of $1/2\xi$ is also indicated

In this comparison of theory and experiment it is important to note that Fig. 8 displays the data in an enlarged fashion. Poisson-Boltzmann theory predicts ϕ to be smaller than 1 and to vary roughly within the range 0.18–0.22. The measured values are located around 0.18. Hence, the dominant change in ϕ, a reduction by a factor of 5, is correctly accounted for. However, on the enlarged scale of Fig. 8 it is evident that the measured values are systematically lower than the prediction, although still higher than the Manning limit $1/2\xi$ that refers to infinite dilution.

Both the correlation-corrected DHHC theory as well as the simulations that capture in principle all kinds of ion correlations (see Sect. 2.2) show a decrease in the osmotic coefficient when compared to the prediction of the PB-theory. Since these two totally different approaches agree so well, it becomes clear that they indeed give a good description of the influence of the correlations. However, they do not lower the osmotic coefficient sufficiently to reach full agreement with the experimental data. Moreover, the deviation from the Poisson-Boltzmann curve increases for higher densities, which is true for the DHHC and the simulations as well as for the experiment. This appears plausible if one recalls that correlations become more important at higher densities.

The fact that Poisson-Boltzmann theory overestimates the osmotic coefficient is well-known in literature. Careful studies of typical flexible polyelectrolytes in solution ([2, 23] and further references given there) indicated that agreement of the Poisson-Boltzmann cell model and experimental data could only be achieved if the charge parameter ξ was renormalized to a higher value. To justify this procedure it was assumed that the flexible polyelectrolytes adopt a locally helical or wiggly main chain in solution. Hence, the counterions "see" more charges per unit length, i.e., a macroion having a higher charge parameter. However, the results obtained for stiff-chain macroions under consideration here [59] show that the osmotic coefficient is lower than the Poisson-Boltzmann results even for systems where the local conformation of the macroion is absolutely rod-like. Evidently, this explanation that has been accepted in literature for a long time [23, 24] is not valid.

Since Poisson-Boltzmann theory neglects all ion-ion correlations (see Sect. 2.2) one is tempted to assume that their incorporation into the theoretical treatment would resolve the discrepancy. However, the comparison displayed in Fig. 8 shows clearly that these correlational effects can only be made responsible for a part of the deviations. Since the two different approaches, using a correlation-corrected density functional theory and Molecular Dynamics simulations, agree very well with each other, it becomes obvious that the discrepancy between them and the experiment is not due to the neglect of ionic correlations.

An important effect not taken into account by the various models discussed in Sect. 2 is the specific interactions of the counterions with the macroion. It is well-known that counterions may even be complexated by macroions and these effects have been discussed abundantly in the early literature in the field [24]. From the above discussion it now becomes clear that these effects must be traced back to specific effects which are not related to the electrostatic interaction of counterions and macroions. Hence, hydrophobic interactions related to subtle changes in the hydratation shell of the counterions could be responsible for this small but significant discrepancy of the electrostatic theory and experiment. Further studies using the PPP-polyelectrolytes will serve for a quantitative understanding of these effects which are outside of the scope of the present review.

4.2.3
Comparison to Data taken from Flexible Polyelectrolytes

Flexible polyelectrolytes that have been studied in solution for decades [23, 24] are out of the scope of the present review. It is interesting, however, to briefly compare the osmotic coefficient obtained from these systems to the data obtained recently from the stiff-chain polyelectrolytes [58]. The osmotic coefficient of Na-polystyrene sulfonate was the subject of a thorough investigation [69]. The experimental procedure applied to measure the osmotic coefficient differed somewhat from the common ones in that the sedimentation equilibrium in an analytical ultracentrifuge was analyzed. In this particular study, polyelectrolytes having a narrow molecular weight distribution, $M_w/M_n<1.1$, were used. The solutions studied had a monomolar concentration of the polyelectrolyte around some 10^{-3} mol/L and contained around 10^{-4} mol/L of a low molecular weight salt. In sedimentation equilibrium, the concentration gradients of both components are coupled via a Donnan-type equilibrium which is governed by the effective charge number of the polyion. Hence ϕ could be obtained from a determination of both concentration gradients.

The data demonstrate that the osmotic coefficient decreases distinctively with rising chain length [69]. The value expected from Manning's limiting law is 0.22, which was observed only for low molar masses up to 200.000 g/mol. For higher molar masses, ϕ drops down to 0.1 at $M_w=10^6$ g/mol [69]. The finding discussed in ref.[69] is supported by other experimental data [70]. Its theoretical basis is still rather unclear, but one can speculate that again the coiling of the polyion may be the underlying cause. A direct comparison to the results obtained on stiff-chain polyions discussed above is not possible since the chain lengths differ tremendously. The contour length of the **PPP-1** samples is around 20 nm, while the shortest polystyrene sulfonates investigated had contour lengths of 100 nm and there are no stiff macromolecules yet in that range. Further measurements done using PPP-polyelectrolytes differing in molecular weight are necessary to come up with a valid comparison with the data of ref. [69].

4.3
SAXS, ASAXS

In recent papers we have shown that small-angle X-ray scattering (SAXS) is a highly suitable method to investigate stiff-chain polyelectrolytes [71]. In particular, it has been demonstrated there that the effect of anomalous dispersion [72] can be applied to discern the contribution of the counterions to the measured scattering intensity $I(q)$. Here the main points of this analysis that is based on earlier work by small-angle neutron scattering (SANS; [73–76]) and by SAXS [77, 78] are presented and discussed.

The scattering intensity $I(q)$ measured for a solution of N polyelectrolyte molecules dispersed in a volume V may be rendered by

$$I(q) = \frac{N}{V} \cdot I_o(q) S(q) \tag{9}$$

where $I_0(q)$ denotes the scattering intensity of a single molecule and $S(q)$ is the effective structure factor to be discussed further below. Since the PPP-polyelectrolyte assume a rod-like shape, the measured scattering intensity $I_0(q)$ can be approximated for higher scattering angles as

$$I_0(q) \to L\frac{\pi}{q} F_{cr}(\Delta\rho(r), q, \alpha=0) F_{cr}^*(\Delta\rho(r), q, \alpha=0) \tag{10}$$

Here $\Delta\rho(r)$ is the radial excess electron density and L is the length of the rod. The quantity $F_{cr}(\Delta\rho(r),q)$ is the Hankel-transform of $\Delta\rho(r)$ whereas α is the cosin of the angle between the scattering vector q and the long axis of the rod. The present analysis therefore rests on the fact that a long rod scatters only radiation if the scattering vector is perpendicular to the long axis of the rod. Hence, scattering experiments probe directly the distribution

$n(r)$ of counterions along the radial direction. The results thus obtained can therefore be directly compared to the PB-cell model.

While SANS allows to vary the contrast between counterions and the macroion [73–76] SAXS usually gives only a combination of the scattering intensity emanating from all parts of the polyelectrolyte. Anomalous small-angle X-ray scattering (ASAXS) provides a solution to this problem inasmuch the anomalous dispersion of the counterions allows to change their X-ray contrast [71]: For many counterions the absorption edge may be reached by synchrotron radiation [72]. The excess electron density of a dissolved ion is given by

$$\Delta\rho_{ion} = \frac{f}{V_{ion}} - \rho_m \tag{11}$$

where V_{ion} denotes the volume occupied by the ion in the solvent and ρ_m is the electron density of the solvent. The scattering factor f may be rendered as

$$f = f_0 + f' + if'' \tag{12}$$

where f_0 is the scattering factor far below the adsorption edge and f' and f'' are the resonant contributions [72]. In the immediate vicinity of the edge f' and f'' are significantly different from zero. Hence, conducting SAXS experiments far below the edge as well as in the neighborhood of the edge allows to assess the contributions of the counterions separately from the contribution of the macroion [71]. First experiments done with polyelectrolyte **PPP-2** demonstrate indeed that ASAXS is a viable method to study dissolved polyelectrolytes up to the region of highest scattering angles [71].

The comparison of the PB-theory and experiment proceeds as follows [71, 73–78]: The radial distribution $n(r)$ of the counterions is calculated according to Eq. (3) of Sect. 2.1. Subsequently the scattering intensity $I_0(q)$ is calculated according to Eq. (10) (see above) or by use of the exact expression [71, 73, 77]. For higher scattering angles the influence of $S(q)$ can safely be dismissed [71] and measured and calculated intensities can directly be compared.

The comparison of measured SAXS data to the predictions of the PB-cell model demonstrate that theory provides a nearly quantitative description of the experimental results. The PB-theory predicts that most counterions are located in the immediate vicinity of the macroions [71]. Hence, the influence of polyelectrolyte concentration is expected to be small because the number of counterions at larger distance to the macroion is small. This is borne out directly from experimental data displayed in Fig. 9. Here the SAXS-intensities of polyelectrolyte **PPP-2** have been normalized to their respective volume fractions and plotted against the magnitude of the scattering vector q. The data superimpose at higher scattering angles but differ at smaller q-values. The latter effect can be traced back to the effect of mutual interaction to be discussed further below. The good agreement at higher scattering angles demonstrates that the mutual interaction of the counterion clouds of the

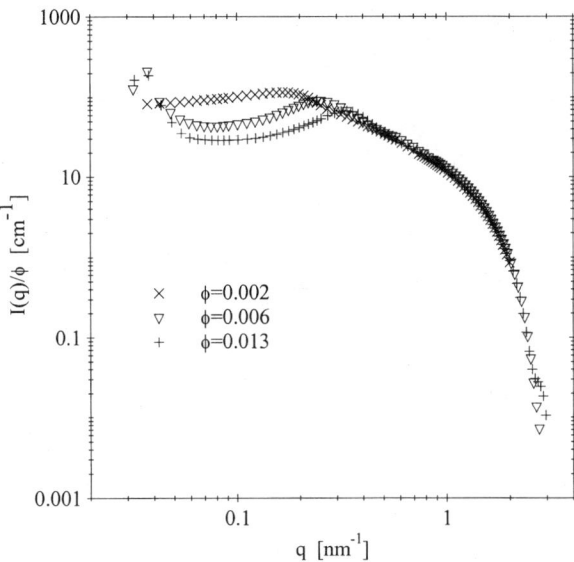

Fig. 9 SAXS-intensities of polyelectrolyte **PPP-2** normalized to the respective volume fractions indicated in the graph [71]. The scattering intensities superimpose for higher q-values but differ at small scattering angles. The latter effect can be traced back to the effect of mutual interaction

rods has a negligible effect on the distribution $n(r)$ as predicted by the PB-theory [71].

Figure 10 furthermore shows that the PB-theory describes the experimental SAXS-data in a virtually quantitative fashion [71]. This is in general accord with previous studies done with monovalent counterions [73–75, 77, 78]. Here the comparison is made for data obtained for two different counterions that differ strongly with regard to contrast. In case of Iodine counterions the contribution of the macroion is rather small. For Chloride counterions, on the other hand, the contributions of the macroion and the counterions are of comparable magnitude. Figure 10 shows that both data can be described satisfactorily by the PB-theory. It must be noted, however, that the analysis of polyelectrolyte **PPP-2** by SAXS has revealed that there is an additional contribution to the measured scattering intensity at high q-values. As discussed in Ref. [71] a possible explanation may be sought in the fluctuations of the counterions along the long axis of the rod-like macroions. Here additional experiments by ASAXS are necessary to come to final conclusions.

Figure 11 serves for the discussion of the region of smallest scattering angles. Here the mutual interaction of the rods leads to a depression of the measured intensity at smallest q. Similar findings have been made in the course of the study of rod-like viruses by Maier et al. [79]. As a consequence, a pronounced maximum of $I(q)$ is visible whereas all scattering curves coin-

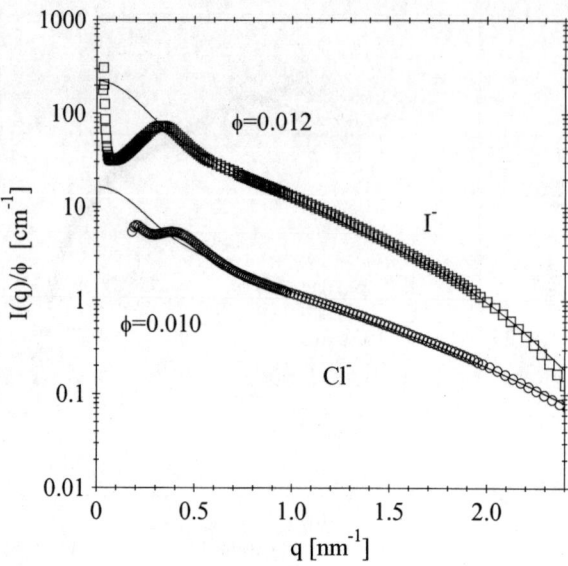

Fig. 10 Comparison of experimental data of polyelectrolyte **PPP-2** to the PB-theory [71]. The respective volume fractions ϕ of the polyelectrolytes are indicated in the graph. The upper curve refers to Iodide counterions whereas the lower curve refers to Chloride counterions

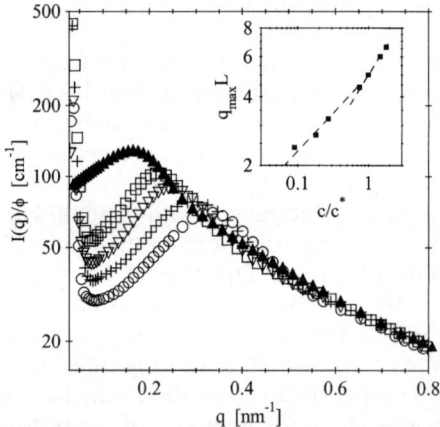

Fig. 11 Interaction of polyelectrolyte rods in salt-free solution: SAXS-intensities measured for different polyelectrolyte concentrations at smallest scattering angles [71]. The respective concentrations are: filled trangles: 3 g/L; hollow triangles 10g/L; crosses: 15 g/L; circles: 20 g/L. The inset displays the maximum of the scattering intensity as function of the reduced concentration c/c^* where $c^*=L^{-3}$

cide for $q>0.6$nm^{-1}. The inset of Fig. 11 displays the position q_{max} of the maximum as a function of the reduced concentration c/c^* where $c^*=L^{-3}$ denotes the overlap concentration. At reduced concentrations smaller than unity q_{max} scales approximately with $(c/c^*)^{1/3}$ as expected for a system of weakly interacting rods [79]. For larger concentrations the exponent 1/2 is found which may be taken as evidence for the onset of mutual alignment of the rods. This finding corroborates a careful study carried out on rod-like viruses [79].

5
Conclusion

A survey over the area of stiff-chain polyelectrolytes has been given. Such rod-like polyelectrolytes can be realized by use of the poly(p-phenylene) backbone [9–13]. The PPP-polyelectrolytes present stable systems that can be studied under a wide variety of conditions. Moreover, electric birefringence demonstrates that these macroions form molecularly disperse solution in water [49]. The rod-like conformation of these macroions allows the direct comparison with the predictions of the Poisson-Boltzmann cell model [27–30] which has been shown to be a rather good approximation for monovalent counterions but which becomes an increasingly poor approximation for higher valent counterions [29]. Here it was shown in Sect. 2.2 that the basic problem of the PB model, namely the neglect of correlations, can be remedied in a systematic fashion.

The osmotic coefficient obtained experimentally from polyelectrolyte **PPP-1** having monovalent counterions compares favorably with the prediction of the PB cell model [58]. The residual differences can be explained only partially by the shortcomings of the PB-theory but must back also to specific interactions between the macroions and the counterions [59]. SAXS and ASAXS applied to **PPP-2** demonstrate that the radial distribution $n(r)$ of the cell model provides a sufficiently good description of experimental data.

Evidently, more work has to be done for a comprehensive comparison of theory and experiment. Theory and simulations reveal clearly that the PB-cell model should be a poor approximation for divalent counterions and breaks down totally for trivalent counterions [29]. A comprehensive experimental test of these very important conclusions is still missing.

Acknowledgment The authors are indebted to the Deutsche Forschungsgemeinschaft for generous support within the "Schwerpunkt Polyelektrolyte". C. Holm acknowledges M. Deserno for intensive collaboration on the presented material.

References

1. Förster S, Schmidt M (1995) Adv Polym Sci 120:51
2. Schmitz KS (1993) Macroions in Solution and Colloid Suspension. VCH, New York
3. Förster S, Schmidt M, Antonietti M (1990) Polymer 31:781
4. MacCallum MJ, Vincent CA (1987) Polymer Electrolyte Reviews, Elsevier, London

5. Barrat J-L, Joanny J-F (1996) Theory of polyelectrolyte solutions. In: Prigogine I, Rice SA (eds) Advances on Chemical Physics 94, Wiley, p 1
6. Kassapidou K, Jesse W, Kuil ME, Lapp A, Egelhaaf S, van der Maarel JRC (1997) Macromolecules 30:2671
7. Raspaud E, da Conceicao M, Livolant F (2000) Phys Rev Lett 84:2533
8. Lee CC, Chu S-G, Berry GC (1983) J Polym Sci Polym Phys Ed 21:1573
9. Metzger Cotts P, Berry GC (1983) J Polym Sci Polym Phys Ed 21:1255
10. Rau IU, Rehahn M (1993) Makromol Chem Phys 194:2225
11. Rau IU, Rehahn M (1993) Polymer 34:2889
12. Rau IU, Rehahn M (1994) Acta Polym 45:3
13. Brodowski G, Horvath A, Ballauff M, Rehahn M (1996) Macromolecules 29:6962
14. Wittemann M, Rehahn M (1998) J Chem Soc Chem Commun 623
15. Rulkens R, Schulze M, Wegner G (1994) Macromol Rapid Commun 15:669
16. Vanhee S, Rulkens R, Lehmann U, Rosenauer C, Schulze M, Köhler W, Wegner G (1996) Macromolecules 29:5136
17. Ballauff M (1993) Mater Sci Tech 12:213
18. Ballauff M (1989) Angew Chem 101:261
19. Rehahn M, Schlüter A-D, Wegner G, Feast WJ (1989) Polymer 30:1060
20. Rehahn M, Schlüter A-D, Wegner G (1990) Makromol Chem Phys 191:1991
21. Schlüter A-D, Wegner G (1993) Acta Polym 44:59
22. Manning GS (1969) J Chem Phys 51:924, 934,3249
23. Oosawa F (1971) Polyelectrolytes, Marcel Dekker, New York
24. Katchalsky A (1971) Pure Appl. Chem. 26:327
25. Deserno M, Ph.D. thesis, Universität Mainz, 2000, http://archimed.uni-mainz.de/pub/2000/0018/;
    Deserno M, Holm C (2001) Cell model and Poisson-Boltzmann theory: a brief introduction. In: Holm C, Kekicheff P, Podgornik R (eds) Electrostatic Effects in Soft Matter and Biophysics. Kluwer, Dordrecht, p 27
26. Deserno M, Holm C, Kremer K (2000) Molecular dynamics simulations of the cylindrical cell model. In: Radeva T (ed) Physical Chemistry of Polyelectrolytes. Marcel Dekker, New York, pp 26
27. Fuoss RM, Katchalsky A, Lifson S (1951) Proc Natl Acad Sci USA 37:579; Alfrey T, Berg PW, Morawetz H (1951) J Polym Sci 7:543
28. Le Bret M, Zimm BH (1984) Biopolymers 23:287
29. Deserno M, May S, Holm C (2000) Macromolecules 33:199
30. Marcus RA (1955) J Chem Phys 23:1057
31. Barbosa MC, Deserno M, Holm C (2000) Europhys Lett 52:80
32. Borukhov I, Andelman D, Orland H (1997) Phys Rev Lett 79:435
33. Lue L, Zoeller N, Blankschtein D (1999) Langmuir 15:3726
34. Nordholm S (1984) Chem Phys Lett 105:302; Penfold R, Nordholm S, Jönsson B, Woodward CE (1990) J Chem Phys 92:1915
35. Groot R (1990) J Chem Phys 95:9191; Diel A, Barbosa MC, Tamashiro MN, Levin Y (1999) Physica A 274:433
36. Gonzales-Tovar E, Lozada-Cassou M, Henderson D (1985) J Chem Phys 83:361; Das T, Bratko D, Bhuyan LB, Outhwaite CW (1997) J Chem Phys 107:9197; Kjellander R (2001) Distribution function theory of electrolytes and electrical double layers. In: Holm C, Kekicheff P, Podgornik R (eds) Electrostatic Effects in Soft Matter and Biophysics. Kluwer, Dordrecht, p. 317
37. Deserno M, Jimenez-Angeles F, Holm C, Lozada-Cassou M (2001) J Phys Chem B, 105:10983; Messina R, Gonzalez Tovar E, Lozada-Cassou M, Holm C (2002) Europhys Lett 60:383
38. Kuhn P, Levin Y, Barbosa MC (1998) Macromolecules 31:8347; Levin Y, Barbosa MC (1997) J Phys II France 7:37
39. Nyquist RM, Ha B-Y, Liu A (1999) Macromolecules 32:3481
40. O'Konski CT (ed) (1978) Molecular Electro-Optics, part 2. Marcel Dekker, New York

41. Jennings BR (ed) (1979) Electro-Optics and Dielectrics of Macromolecules and Colloids. Plenum Press, New York
42. Krause S (ed) (1981) Molecular Electro-Optics. Plenum Press, New York
43. Fredericq E, Houssier C (1973) Electric Dichroism and Electric Birefringence. Clarendon Press, Oxford
44. Yamaoka K, Ueda K (1980) J Phys Chem 84:1422.
45. Schwarz G (1959) Z Phys 145:563; Z Phys Chem 19:286
46. Mandel M (1961) Mol Phys 4:489.
47. Oosawa F (1970) Biopolymers 9:677
48. Lachenmayer K (2000) PhD. Thesis, Stuttgart.
49. Lachenmayer K, Oppermann W (2002) J Chem Phys 116:392.
50. O'Konski CT, Krause S (1970) J Phys Chem 74:3243.
51. van Dijk W, van der Touw F, Mandel M (1981) Macromolecules 14:792.
52. Hogan M, Dattagupta N, Crothers DM (1978) Biochemistry 75:195
53. Rau DC, Bloomfield VA (1979) Biopolymers 18:2783
54. Fixman M, Jagannathan S (1981) J Chem Phys 75:4048
55. Yoshida M, Kikuchi K, Maekawa T, Watanabe H (1992) J Phys Chem 96:2365
56. Oppermann W (1988) Macromol Chem 189:927; 189:2125
57. Wandrey Ch, Hunkeler D, Wendler U, Jaeger W (2000) Macromolecules 33:7136
58. Blaul J, Wittemann M, Ballauff M, Rehahn M (2000) J Phys Chem B 104:7077
59. Deserno M, Holm C, Blaul J, Ballauff M, Rehahn M (2001) Eur Phys J E 5:97
60. Deserno M, Holm C (2002) Molecular Physics 100:2941
61. Neu J (1999) Phys Rev Lett 82:1072
62. Sader J, Chan DY (1999) J Colloid Interface Sci 213:268
63. Trizac E, Raimbault JL (1999) Phys Rev E 60:6530
64. Nilsson LG, Guldbrand L, Nordenskiöld L (1991) Mol Phys 72:177
65. Lyubartsev AP, Nordenskiöld L (1997) J Phys Chem 101:4335
66. Gronbech-Jensen N, Mashl RJ, Bruinsma RF, Gelbart WM (1997) Phys Rev Lett 78:2477
67. Lyubartsev AP, Tang JX, Janmey PA, Nordenskiöld L (1998) Phys Rev Lett 81:5465
68. Auer H, Alexandrowicz Z (1969) Biopolymers 8:1
69. Oppermann W, Wagner M (1999) Langmuir 15:4089
70. Wandrey C (1997) Polyelektrolyte—Makromolekulare Parameter und Elektrolytverhalten. Cuvillier, Göttingen
71. Guilleaume B, Blaul J, Wittemann M, Rehahn M, Ballauff M (2002) Eur Phys J E 8:299; Guilleaume B, Ballauff M, Goerigk G, Wittemann M, Rehahn M (2001) Colloid Polym Sci 279:829
72. Stuhrmann HB Adv Polym Sci (1985) 67:123
73. Kassapidou K, Jesse W, Kuil M E, Lapp A, Egelhaaf S, van der Maarel JRC (1997) Macromolecules 30:2671
74. van der Maarel JRC, Kassapidou K (1998) Macromolecules 31:5734
75. van der Maarel JRC, Groot LCA, Mandel M, Jesse W, Jannink G, Rodriguez V (1992) J Phys II France 2:109
76. Zakharova SS, Engelhaaf SU, Bhuiyan LB, Outhwaite CW, Bratko D, van der Maarel, JRC (1999) J Chem Phys 111:10706
77. Wu CF, Chen SH, Shih LB, Lin JS (1988) Phys Rev Lett 61:645
78. Chang SL, Chen SH, Rill RL, Lin JS (1990) J Phys Chem 94:8025
79. Maier EE, Krause R, Deggelmann M, Hagenbüchle MM, Weber R, Fraden S (1992) Macromolecules 25:1125

Received: November 2002

Conformation and Phase Diagrams of Flexible Polyelectrolytes

N. Volk[1] · D. Vollmer[1,2] · M. Schmidt[1] · W. Oppermann[3] · K. Huber[4]

[1] Institut für Physikalische Chemie, Johannes Gutenberg-Universität,
55099 Mainz, Germany
E-mail: volk@mail.uni-mainz.de
E-mail: vollmerd@mail.uni-mainz.de
E-mail: mschmidt@mail.uni-mainz.de
[2] School of Physics, University of Edinburgh, Kings Building, Edinburgh, UK
[3] Institut für Physikalische Chemie der Technischen Universität Clausthal,
38678 Clausthal-Zellerfeld, Germany
E-mail: wilhelm.oppermann@tu-clausthal.de
[4] Institut für Physikalische Chemie, Universität Paderborn, 33098 Paderborn, Germany
E-mail: huber@fb13n.uni-paderborn.de

Abstract The present article addresses the dilute solution behavior of two different polyelectrolyte systems: Aqueous solutions of quaternized poly(2-vinylpyridines) and of polyacrylic acids. Firstly, it is demonstrated that the dimensions of the chains for all investigated polyelectrolytes are described by a model that explicitly considers (i) an excluded volume comprising contributions of the electrostatic interactions via the effective charge density, and (ii) the intrinsic excluded volume in terms of the Flory interaction parameter. The effect of the chain hydrophobicity and the type of counterions on the coil dimension of the chains and the effective charge density is discussed. The latter is compared with results obtained from osmotic pressure measurements and conductometry.
 The second part of the review is devoted to an investigation of the phase behavior of the two systems. For quaternized poly-2-vinylpyridines phase transitions are induced by addition of inert salt, typically NaI and are denoted as "salting out" and "salting in". The phase behavior is assumed not to be influenced by ion specific interactions other than the Flory excluded volume parameter. Contrary, in case of polyacrylic acid highly specific interactions, such as complexation, between the polyion and bivalent earth alkaline cations may cause a precipitation of the polyelectrolyte. The precipitation point depends on both, the concentration of the polyions and the counter ions. Investigation of the coil dimensions by means of combined static and dynamic light scattering reveals a coil collapse towards spherical particles. Possible transition states along this shrinking process are discussed.

Keywords Polyelectrolytes · Manning condensation · Phase diagrams of polyelectrolytes · Polyacrylic acids · Poly(2-vinylpyridine) · Osmotic coefficient · Light scattering · Conductometry

1	Introduction.................................	30
2	Dilute Solution Behavior of Sodium Polyacrylate and Quaternized Polyvinylpyridine.................	32
2.1	Theory.......................................	32
2.1.1	Effective Charge Density and Counterion Condensation........	32
2.1.2	Polyelectrolyte Dimensions	33

© Springer-Verlag Berlin Heidelberg 2004

2.2   Dilute Solution Properties of Polyvinylpyridinium Cations   36
2.2.1 The Effect of Hydrophobicity   37
2.2.2 The Effect of Counterions   39
2.3   Dilute Solution Properties of Sodium Polyacrylate.   40
2.4   Effective Charge Density at Low and High Salt Concentrations . .   43
2.4.1 Effective Charge Density Determined by the Osmotic Coefficient.   43
2.4.2 Conductivity Measurements Across the Phase Boundary   46
2.4.3 Discussion of the Effective Charge Densities   49

3   **Aspects of Phase Behavior of Quaternized Polyvinylpyridine
       and Sodium Polyacrylate**.   50

3.1   Salting Out and Salting in of Poly(2-vinylpyridinium) Cations. . .   50
3.2   L-type Precipitation of Sodium Polyacrylate   54

4   **Size and Shape of Collapsing Polyelectrolytes by Light Scattering**   58

4.1 Ca^{2+} Induced L-Type Precipitation of Sodium Polyacrylate   58
4.2   Comparison of Ca^{2+}, Sr^{2+} and Ba^{2+} Ions   62

References   64

Abbreviations

DMF	dimethylformamide
GPC	gel permeation chromatography
LS	light scattering
NaPA	sodium salt of polyacrylic acid
NaPMA	sodium salt of polymethacrylic acid
NaPSS	sodium salt of polystyrene sulfonic acid
PEO	polyethyleneoxide
PNIPAM	poly(n-isopropylacrylamide)
PVP	poly(2-vinylpyridine)
THF	tetrahydrofuran
SAXS	small angle x-ray scattering

1
Introduction

The behavior of flexible polymers is satisfactorily described by two parameters [1, 2], the excluded volume integral β and the statistical or Kuhn segment length l_k, which may also be expressed in terms of the persistence length l_p. Whereas the latter captures chain stiffness inferred by successive monomers, the excluded volume accounts for intermolecular interactions as well as chain expansion.

The phase separation in solutions of neutral polymers is quite familiar to polymer scientists. It is usually induced by a temperature change which modifies inter- and intramolecular interactions. If the concentration of macromolecules is low enough, changes are restricted to intramolecular interactions and neutral polymers collapse to a more or less swollen globule [3–8]. Unlike that of neutral polymers, the mechanism of coil shrinking for polyelectrolytes and the phase behavior is much less understood.

Polyelectrolyte behavior is governed by electrostatic forces, which by far exceed conventional excluded volume interactions. Meanwhile, extensive knowledge has been collected on dilute solution behavior of polyelectrolytes in aqueous solution. In salt free water, dimensions are highly extended, although still far from adopting a rigid rod like shape [9, 10]. However, intrinsic viscosity together with light and X-ray scattering clearly indicated, that long-range electrostatic interactions lead to clustering of polyelectrolytes, impairing a proper evaluation of the single chain behavior [11–17].

Only if an inert electrolyte like NaCl is added, the strong *electrostatic interactions are increasingly screened* and the highly expanded polyelectrolyte coils start to shrink. Eventually the unperturbed dimensions are approached at high enough concentrations of inert salt [18–20]. In such cases, phase separation occurs if this inert salt level is surpassed [21–23]. The latter phenomena has been denoted as "salting out" of polyelectrolytes, or alternatively, as H-type precipitation, because the concentration of inert salt required to cause precipitation of the polyelectrolyte is high and independent of the polymer concentration [23].

This rather simple scheme of solution behavior is complicated by various interesting aspects. It is not at all clear that the number of dissociated or "non-condensed" counterions remain constant with added salt. Likewise, at high salt concentrations, the polyelectrolyte solution may no longer represent a quasi binary solution but rather a ternary mixture with the added salt representing a third component with distinctly different solvation properties as compared to water. This scenario is supported by the fact that for some salts the ordinary salting out is reversed if the inert monovalent salt concentration is further increased. In other words the chains become redissolved at an even higher salt concentration [21, 22]. Consequently, this redissolution is denoted as "salting in".

In principle multivalent counterions produce a qualitatively similar behavior as long as no specific interactions, such as complexation, occur between the polyion and the counterions. However, a completely different phase behavior is observed if *specific interactions* between the counterions and the polyion charge exist, which is frequently found for multivalent counterions. In such cases, even a much lower salt concentration may cause a precipitation of the polyelectrolyte. This kind of phase separation is denoted as L-type precipitation [23–25]. For example, the amount of earth alkaline cations required to precipitate polyacrylate anions increases with increasing polyanion content, being roughly equimolar if no additional inert salt is present. Addition of an inert salt shifts this phase boundary towards larger concentrations of bivalent metal cations. As in the H-type precipitation, redissolution of the precipitate may occur. Two different origins for this redissolution are discussed in literature. One possible origin is that a continuous increase of the concentration of multivalent metal cations screens attractive interactions between oppositely charged backbone sections. Those include positively charged sections due to the complexation of multivalent cations and fully dissociated, negatively charged sections [26]. Alternatively, a dominating monodentate binding of multivalent cations may surpass the point of electro-neutrality and leads to a charge inversion of the polyelectrolyte chains, which then become redissolvable [27].

The present review deals mainly with two examples of polyelectrolyte phase behavior as discussed above. As an example for an H-type precipitation, the solution properties of polyvinylpyridinium chains are monitored as function of added inert salt. Here, we focus on the determination of the effective charge density and of the solvent quality parameter which are supposed to play a central role for the understanding of polyelectrolyte solution without specific counterion interactions. The second system under investigation comprises the interaction of polyacrylic acid with alkaline earth cations which exhibit very specific interactions, thus representing an example for type L-precipitation. Here the coil dimensions close to the phase boundary are compared to those close to type H-precipitation with inert added salt.

2
Dilute Solution Behavior of Sodium Polyacrylate and Quaternized Polyvinylpyridine

2.1
Theory

2.1.1
Effective Charge Density and Counterion Condensation

Polyelectrolyte theories generally make use of the concept of counterion condensation [28–31]. This means that strongly charged polyelectrolytes reduce their line charge density by having a certain fraction of the counterions tightly bound (condensed) to the polyion. The decisive parameter is the normalized charge density $\xi = l_B/a$ with a the average linear distance between charges. l_B is the Bjerrum length

$$l_B = \frac{e^2}{4\pi\varepsilon_r\varepsilon_0 k_B T} \tag{1}$$

where e is the elementary charge, ε_r is the effective dielectric constant of the solution, ε_0 is the permittivity of the vacuum, and $k_B T$ is the Boltzmann constant times the absolute temperature. According to Manning's limiting law [29–31], monovalent counterions will condense on the polyion when $\xi > 1$, and the amount of condensed counterions is such as to reduce the effective charge density to 1. The uncondensed mobile ions are subject to electrostatic interaction with the polyion.

For a more detailed discussion of this point the reader is referred to other contributions in this volume [Holm et al., Ballauff et. al].

2.1.2
Polyelectrolyte Dimensions

Theories of conformations of polyelectrolytes fall into two groups. In the first group [32–34] the chain is assumed to be a flexible chain and the consequence of electrostatic interaction is calculated. In the second category [35–42], the chain is assumed to be a stiff chain and calculations are performed to obtain the effect of the electrostatic interaction between charges on the chain backbone. To date, there is no satisfactory theory in the literature to describe the electrostatic effect on conformations of polyelectrolyte chains with arbitrary intrinsic stiffness. In the following we briefly outline the developments for both groups of theories.

In the first approach the effect of electrostatic interaction on a flexible polyelectrolyte chain in infinitely dilute solution has been modeled by the following parameters. The chain contains N_k Kuhn segments with step length l_k. The total charge on the chain, $l_k N_k eZf/b$ is assumed to be uniformly distributed along the chain backbone, where f is the effective charge density, b the monomer length and Z is the valence of the ionic group. The interaction energy $V(r)$ between the segments separated by a distance r is taken as

$$\frac{V(r)}{k_B T} = w \cdot \delta(r) + \frac{Z^2 f^2 l_B e^{-\kappa r}}{r} \qquad (2)$$

where the first and second terms on the right hand side describe the usual short-ranged excluded volume effect and the electrostatic interaction respectively. w is the pseudo-potential and is equivalent to $(0.5-\chi)l_k^3$ with χ being the Flory-Huggins interaction parameter. $\delta(r)$ is the Dirac delta function and κ is the inverse Debye length given by

$$\kappa^2 = 4\pi l_B \left(Z_c^2 c_c + \sum_\gamma Z_\gamma^2 c_\gamma \right) \qquad (3)$$

where Z_c and c_c are respectively the valence and the number concentration of the counterion to the polyelectrolyte. Z_γ and c_γ are the valence and number concentration of γth salt ion.

Utilizing the intermolecular interaction potential described by Eq. (2) and assuming a uniform expansion for the chain the expansion factor α for the radius of gyration

$$\alpha = \frac{R_g}{R_{g,\Theta}} \qquad (4)$$

with $R_{g,\Theta}$ the unperturbed radius of gyration, was derived by a variational procedure [33, 34].

In the limit of weak interactions, i.e. $\kappa^2 l_{k^2} N_k \alpha^2 / 6 \gg 1$, the following simple equation is obtained

$$\alpha^5 - \alpha^3 = \left[w_0 + \frac{134}{35} \sqrt{\frac{6 l_B Z^2 f^2}{\pi l_k \kappa^2 b^2}} \right] \sqrt{N_k} \qquad (5)$$

where

$$w_0 = \frac{134}{105}\left(\frac{3}{2\pi}\right)^{3/2}\left(\frac{1}{2}-\chi\right) \tag{a}$$

Equation 5 corrects the expression given by Beer et al. [43] by a factor l_k^2/b^2. It is equivalently written in the form of the modified Flory formula [1], as

$$\alpha^5 - \alpha^3 = \frac{134}{105}\left(\frac{3}{2\pi}\right)^{3/2}\left[\left(\frac{1}{2}-\chi\right) + 4\pi\frac{l_B Z^2 f^2}{\kappa^2 l_k^3}\right]\sqrt{N_k} \tag{6}$$

This result can readily be guessed directly from the equation for $V(r)$. If κ is large, the second term on the right hand side of Eq. (2) becomes short-ranged, $4\pi Z^2 f^2 l_B \kappa^{-2}\delta(r)$ and the prefactor of $N_k^{0.5}$ in Eq. (5) reducues to w_0. This point was noted many years ago by Flory [32].

For theta solutions ($w_0=0$) and for $N_k \gg 1$ and $\kappa l_k^2 N_k \alpha^2/6 \ll 1$, α can be derived to be

$$\alpha = \left(\frac{134}{1575}\right)^{3/2}\left(\frac{6}{\pi}\right)^{1/6}\left(\frac{Z^2 f^2 l_k}{b^2}\right)^{1/3}\sqrt{N_k} \tag{7}$$

so that the radius of gyration R_g follows as

$$R_g = \left(\frac{134}{1575}\right)^{1/3}\left(\frac{6}{\pi}\right)^{1/6}\frac{1}{\sqrt{6}}\left(\frac{Z^2 f^2 l_k^5 l_B}{b^2}\right)^{1/3} N_k \tag{8}$$

with the same scaling form as originally obtained by Kuhn, Künzle and Katchalsky [44], which in turn has recently been interpreted, using the electrostatic blob model [45].

Since several approaches exist to determine the effective charge density, we denote f determined from a fit of R_g as f_α throughout the paper.

The alternative method [46] of analyzing the data is based on the treatment by Odijk and Houwaart [36] of the excluded volume effect on the electrostatic stiffening of semi flexible chains. The total persistence length l_t of a stiff polyelectrolyte is the sum of the intrinsic persistence length $l_p \equiv l_k/2$ and the electrostatic persistence length l_e,

$$l_t = l_p + l_e \tag{9}$$

As shown by Odijk [35] and Skolnick and Fixman [37],

$$l_e = \frac{1}{4 l_B \kappa^2} \tag{10}$$

in the limit of $\kappa l_k \sqrt{N_k} \gg 1$ and $l_e < l_p$. In addition to the electrostatic persistence length generated by the charge on the chain backbone, Odijk and Houwaart [36] assume further that there is an excluded volume interaction among cylindrical segments each of length $2l_t$ and diameter $4\kappa^{-1}$. In general, the excluded volume parameter for a flexible chain of Kuhn segments with Kuhn length l_k is

$$\tilde{z} = \left(\frac{3}{2\pi l_k^2}\right)^{3/2} \beta \sqrt{N_k} \tag{11}$$

where β is the excluded volume. Several approximate formulas for the dependence of the expansion factor α on $\tilde{z}$ have been reported in the literature [1, 47, 48]. For the specific model of Odijk and Houwaart, β is solely electrostatic in nature and is given as

$$\beta = \beta_{el} = 8\pi \kappa^{-1} l_t^2 \tag{12}$$

Combining with the Yamakawa-Tanaka relation [1] for α in terms of $\tilde{z}$,

$$\alpha^2 = 0.541 + 0.459(1 + 6.04\tilde{z})^{0.46} \tag{13}$$

Eq (11) and (12) lead to the result

$$R_g^2 = \frac{L_w}{3}\left[l_p + \frac{1}{4l_B\kappa^2}\right]\left(0.541 + 0.459\left\{1 + 4.43 L_w^{1/2}\kappa^{-1}\left[l_p + \frac{1}{4l_B\kappa^2}\right]^{-3/2}\right\}^{0.46}\right) \tag{14}$$

Here L_w is the contour length of the chain. In an effort to generalize Eq. (14) for non-theta solutions, Reed et al. [46] have provided an ad hoc treatment by combining theories of Odijk [35], Odijk and Houwaart [36], Skolnick and Fixman [37], and Gupta and Forsman [49]. Here they append an additional contribution β_0 to β, arising from short-ranged non-electrostatic interactions, so that β of Eq. (11) is given by

$$\beta = \beta_{el} + \beta_0 \tag{15}$$

where β_0 is taken to be an adjustable parameter (β_0 actually being $0.5-\chi$). In addition, Reed et al. [46] use the empirical formula of Gupta and Forsman [49] (based on Monte Carlo results)

$$\alpha^5 - \alpha^3 = \frac{134}{105}(1 - 0.885 N_k^{-0.462})\tilde{z} \tag{16}$$

instead of the Yamakawa-Tanaka formula, in evaluating the expansion factor. It must be noted that the applicability of Eq. (9) to (16) is constrained by the basic assumption that the electrostatic persistence length l_e is smaller than the intrinsic persistence length l_p. Therefore care must be exercised in comparing the experimental data with the predictions of Eq. (9) to (16) to make sure that the salt concentration is high enough so that $l_e < l_p$.

2.2
Dilute Solution Properties of Polyvinylpyridinium Cations

The preparation and characterization of the samples are described elsewhere [43]. The results are summarized in Tables 1 and 2. The number at the end of the sample code represents the contour length derived from the molar mass of the polymers. The PVP samples were quaternized with the respective alkyl and benzyl bromides in order to yield polypyridinium cations with

Table 1 Characterization of the poly-2-vinylpyridines

Sample	M_w (THF/MeOH)/gmol^{-1}	L_w[nm]	M_w/M_n*	R_g[nm]	R_h[nm]
PVP400	1.68*×10^5	400	1.13	15.6	9.9
PVP690	2.9*×10^5/2.8*×10^5	690	1.24	17	14.5
PVP1000	4.2*×10^5	1000	1.11	25.4	16.6
PVP1800	7.6*×10^5/7.5*×10^5	1800	1.20	37.5	26
PVP2100	8.8*×10^5	2100	1.14	39	27

* Determined by GPC in DMF, polystyrene calibration

Table 2 Degree of ionization Q determined by elemental analysis and by IR-spectroscopy

	C%	H%	N%	Cl, Br%	Q(EA)^a%	Q(IR)^b%
Bz-PVP2100Br	–	–	–	–	–	77
Bz-PVP1800Br	58.3	5.25	5.72	31.0	95	86
Bz-PVP1000Br	57.2	5.50	5.48	31.7	101	91
Bz-PVP690Br	58.0	5.96	5.71	30.3	93	86
Bz-PVP400Br	58.8	5.42	5.56	30.0	94	96
Et-PVP1800Cl	67.7	7.53	9.63	15.6	64	67
Et-PVP1000Cl	67.9	7.34	10.4	14.4	55	58
Et-PVP400Br	56.6	5.98	8.36	29.4	61	58
Bu-PVP1800Br	53.7	6.08	7.1	32.6	80	67
Oc-PVP1800Br	58.2	7.02	6.33	28.5	79	91
Me-PVP1940I	–	–	–	–	–	80
Me-PVP1940I	–	–	–	–	–	50
Et-PVP1940I	–	–	–	–	–	80
Et-PVP1940I	–	–	–	–	–	70
Et-PVP1940I	–	–	–	–	–	40
Bu-PVP1940I	–	–	–	–	–	70
Bu-PVP1940I	–	–	–	–	–	40
ME1BPVP1940Tos^c	–	–	–	–	–	70

^{a} Degree of ionization determined by elemental analysis via the N to Br or Cl ratio
^{b} Determined by IR-spectroscopy
Following quaternizing group are used (Me=methyl, Bz=benzyl, Et=ethyl, Bu=butyl, Oc=octyl, H=hydrogen). Q is the degree of quaternization and Br, Cl and I stand for the elemental code of the counterion. Hydrogenated PVP were prepared by dissolving PVP in aqueous HBr solution at PH=3 which yields $Q \cong 30\%$
^{c} ME1B stand for $CH_3OCH_2CH_2O(CH_2)_4$

different hydrophobicity (see Scheme 1). Different counterions i.e. F⁻, Cl⁻ and I⁻ were introduced by ultra filtration [43]. The degree of chemical ionization was determined by IR-spectroscopy and is given in Table 2 along with the sample codes.

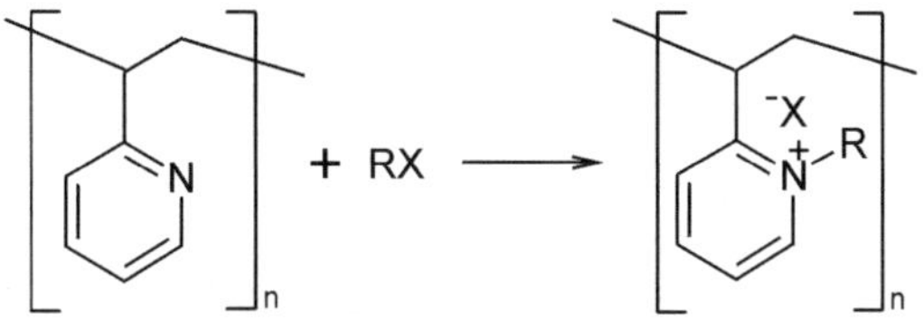

Scheme 1 Sketch of an unquaternized and a quaternized repeat unit of poly(2-vinyl-pyridine)

2.2.1
The Effect of Hydrophobicity

The ionic strength dependence of the polyelectrolyte dimensions for different molar masses are shown in Figs. 1, 2, 3 for the H-PVP, Et-PVP and Bz-PVP samples, respectively. In all these measurements the counterion is Br⁻ and the added salt is NaBr. It is qualitatively recognized that almost all data sets exhibit a constant slope at small c_s irrespective of the hydrophobicity of the polyelectrolyte. The exponents observed vary between −0.15 and −0.24 with the tendency that the slope increases with decreasing hydrophobicity and increasing chain length. The slopes are somewhat smaller than reported for NaPSS [46]. In contrast, at high salt concentration the hydrophobicity becomes the dominating parameter which determines the curvature of the plots and the solubility limit ("salting out effect"). In terms of the theoretical expression given by Eq. (5) the charge density f_α dominates at low c_s whereas

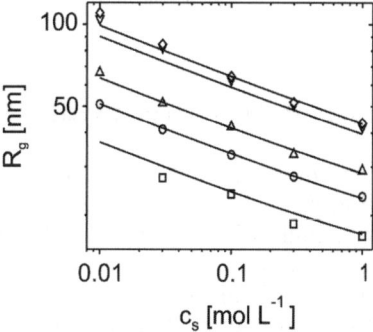

Fig. 1 Ionic strength dependence of the radius of gyration for sample H-PVPL$_w$Br in aqueous NaBr solution: L_w=400 nm (□), L_w=690 nm (○); L_w=1000 nm (△); L_w=1800 nm (▽); L_w=2100 nm (◊). The *solid lines* correspond to the best fit to all data by the theoretical Eq. (5) with f_α=0.22 and w_0=−0.02

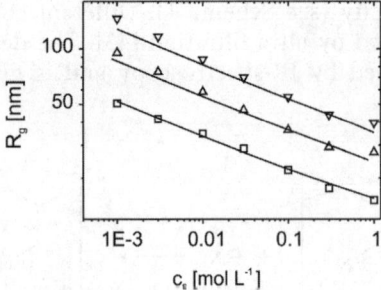

Fig. 2 Ionic strength dependence of the radius of gyration for sample Et-PVPL$_w$Br in aqueous NaBr solution: L_w=400 nm (□); L_w=1000 nm (△); L_w=1800 nm (▽). The *solid lines* are calculated with f_α=0.17 and w_0=-0.03

w_0 essentially controls curvature and precipitation point at high c_s. The theoretical curves shown in Figs. 1–3 almost perfectly fit the experimental points.

The values for f_α and w_0 derived from the theoretical curves (solid lines in

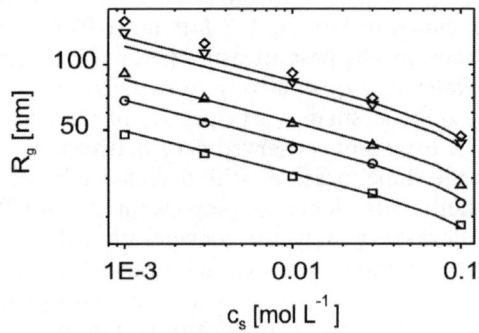

Fig. 3 Ionic strength dependence of the radius of gyration for sample Bz-PVPL$_w$Br in aqueous NaBr solution: *Symbols* as in Fig. 1. The *solid lines* are calculated with f_α=0.15 and w_0=−0.32

Figs. 1–3) are summarized in Table 3 and display following qualitative features:

The value of w_0 for the most hydrophilic H-PVP samples is zero or slightly below zero indicating that the H-PVP chains should adopt an "unperturbed" conformation at high salt concentration. The slightly more hydrophobic Et-PVP exhibits a slightly smaller w_0-value, whereas Bz-PVP with w_0=−0.32 represents the most hydrophobic polyelectrolyte. It is to be noted that w_0=−0.32 leads to precipitation at c_s=0.13 M in good agreement with the experimentally observed solubility limit at c_s=0.14 M (see Table 3).

It is most interesting to note that the fitted value for the effective charge density f_α does significantly depend on the hydrophobicity of the chain and

Table 3 Fit parameters w_0 and f_α for differently quaternized polyvinylpyridines and contour length L_w. The calculated (according to Eq. 5) and the observed "salting out" concentrations, $c_{s,p}$ are also given

	f_α	w_0	$c_{s,p}$ (theo)	$c_{s,p}$ (exp)
H-PVP L_wBr	0.22	−0.02	4.5 m	–
Et-PVP L_wBr	0.17	−0.03	1.8 m	1–2 m*
Bz-PVP L_w Br	0.15	−0.32	0.13 m	0.14 m

* Minimum of R_g vs c_s as discussed in the text

decreases from f_α=0.22 for H-PVP to f_α=0.15 for Bz-PVP. Thus, all fitted charge densities lie well below the Manning counterion condensation limit of f_α=0.35.

2.2.2
The Effect of Counterions

The effect of different counterions on the polyion dimensions is shown in Fig. 4, for the Et-PVP sample with L_w=10² nm. Unexpectedly, a huge effect on f_α and on w_0 is observed when the counterion is changed from F⁻ to Cl⁻ to Br⁻ to I⁻. The curves for the F⁻, Cl⁻ and I⁻ - counterions are well fitted by the f_α and w_0 values listed in Table 4. Only for the Br⁻ counterion a significant deviation is observed at high salt concentrations which allows w_0 to vary between w_0=−0.012 (filled curve for $\triangle$ in Fig. 4) and w_0=0 (dotted curve). The origin of this discrepancy becomes clear if the radii are monitored at even higher salt concentrations as shown in Fig. 5. For NaBr the dimensions start to increase again at higher c_s. For NaI the polyelectrolyte which is salted out at c_s=0.15 M NaI redissolves for c_s>2 M NaI. Now the di-

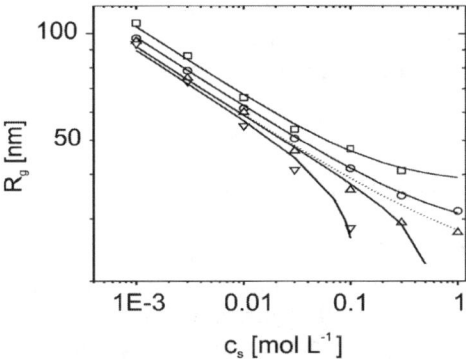

Fig. 4 Ionic strength dependence of the radius of gyration for sample Et-PVP1000X with different counterions: X⁻=F⁻ ($\square$), Cl⁻ ($\bigcirc$), Br⁻ ($\triangle$) and I⁻ ($\triangledown$). The inert salt was NaF, NaCl, NaBr and NaI, respectively. The *solid lines* show least square fits of the data using Eq. (5)

Table 4 Fitted values for f_α and w_0 for the sample Bz-PVP1000X with different counterions X$^-$

X$^-$	f_α	w_0
F$^-$	0.25	+0.5
Cl$^-$	0.21	0.0
Br$^-$	0.18	−0.012/0.0
I$^-$	0.17	−0.5

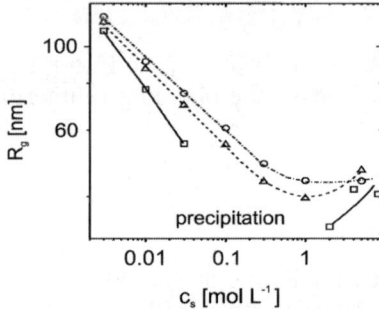

Fig. 5 Ionic strength dependence of the radius of gyration for sample Et-PVP1800X with different counterions: X$^-$=Cl$^-$ (○), Br$^-$ (△) and I$^-$ (□). The inert salt was NaCl, NaBr and NaI, respectively. The *solid lines* show least square fits of the data using Eq. (5)

mensions increase with increasing NaI content. This is a strong indication that w_0 might become a function of added salt for some systems. This effect, however, is expected to have a significant influence at sufficiently high weight fractions of added salt, only. At 1 M salt, the mass fraction of salt is in the order of 10%. Since this effect cannot be quantified it is not accounted for in the data analysis given above. Likewise, the effective charge density may change as function of added salt. This point is addressed in the next section. It should also be noted that the presented data could also be quantitatively fitted by the modified Odjik, Fixman model, i.e. by Eq. (14) to (16). However, the intrinsic persistence lengths, which resulted from the fit seem to be too large.

2.3
Dilute Solution Properties of Sodium Polyacrylate

As early as 1951, Flory and Osterheld [18] could show that partially ionized polyacrylic acid in aqueous solution of an inert salt shrinks in size if the concentration of the inert salt is increased. This shrinking process can be pushed towards the unperturbed dimensions of the NaPA chains. Known from neutral polymers as Θ-state, it is reached for fully ionized NaPA [50] at T=15 °C and 1.5 M KBr. In several papers, the dependence of the intrinsic viscosity was investigated as a function of the molar mass [51]. Data were

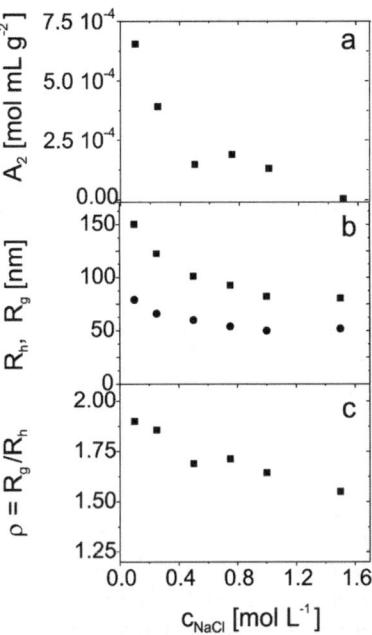

Fig. 6 Dilute solution properties of sample PA2 (M_w=3.3 10^6gmol^{-1}) as a function of the inert salt concentration: **a** Second virial coefficient; **b** radius of gyration (■) and hydrodynamic radius (●); **c** structure sensitive ratio ρ

interpreted in terms of the Yamakawa-Fujji theory [52] of cylindrical worm-like chains. Interpretation resulted in an increase of the persistence length with decreasing inert salt concentration. A more recent light scattering investigation compared radii of gyration and hydrodynamic radii for various NaPA samples at an inert salt level of 1.0 M NaCl with molecular dynamics simulation [53].

In the present review article, we report on molecular weight dependent dynamic and static light scattering data of marginally polydisperse NaPA samples (M_w/M_n=1.2) under two different solvent conditions [54]. The mean square radius of gyration R_g, the second virial coefficient A_2 and the hydrodynamically effective radius R_h was determined in a molecular weight range of 10^5g/mol $<M_w<$3.3·10^6g/mol. This range is wide enough in order to give access to the scaling behaviour of R_g and R_h versus M_w.

In a first step, appropriate solvent conditions were established. This was achieved by investigating a single sample at M_w=3.3×10^6 g/mol as a function of the NaCl content. The latter acts as an inert salt and screens the electrostatic interactions. Results are summarized in Fig. 6.

The second virial coefficient decreases from a value of A_2=6.5 10^{-3} mol·mL/g^2 at 0.1 M NaCl to a value of A_2~0 at 1.5 M NaCl. At the same time, the coil dimensions decreased by some 50%. Measurement of both radii allowed to calculate the shape sensitive ratio

$$\rho = \frac{R_g}{R_h} \tag{17}$$

which, in the present case, decreased from 1.95 in 0.1 M NaCl to 1.6 in 1.5 M NaCl. For neutral polymers in a good solvent, Akcasu and Benmouna [55] predicted a value of 1.86. A slightly lower value of 1.64 could be inferred from Barrett's calculation [56] of the geometric and hydrodynamic expansion coefficients [57]. For neutral chains under Θ-conditions, a significantly lower value of $\rho=1.5$ was evaluated [58]. Experiments with monodisperse neutral polymers confirmed this solvent quality dependent drop in ρ, but the experimental values were generally a few percent lower than theoretically expected [57, 59, 60]. Unlike those of neutral polymers, our results on NaPA seem to be a few percent larger than those theoretically expected. Nevertheless, the drop of ρ observed for NaPA indicates that the NaPA chains undergo a transformation from a self avoiding walk in 0.1 M NaCl (good solvent conditions) to a random walk in 1.5 M NaCL (Θ-conditions), in agreement with the disappearance of the second virial coefficient.

Based on these indications, 0.1 M NaCl and 1.5 M NaCl were selected for measurement of the molar mass dependences of the light scattering parameters under two limiting solvent conditions respectively. The following scaling laws were found:

In 0.1 M NaCl:

$$R_g = 0.0214 \, M_w^{0.60}$$

$$R_h = 0.0112 \, M_w^{0.60} \tag{18a}$$

In 1.5 M NaCl:

$$R_g = 0.0374 \, M_w^{0.52}$$

$$R_h = 0.0232 \, M_w^{0.52} \tag{18b}$$

Clearly, the exponents offered further support for the assumed solvent conditions at 0.1 M NaCl and 1.5 M NaCl. Together with the results from the inert salt dependent experiments shown in Fig. 6, we were able to establish changes of size and shape of the NaPA chains which were induced by regular electrostatic screening effects. Now an important prerequisite has been met in order to successfully isolate and assess specific effects of multi-valent cations on these polyelctrolyte features in the presence of an additional inert salt. This will be discussed in Sect. 3.2.

Beyond that, we succeeded in extracting unperturbed dimensions of NaPA chains under both solvent conditions. Expressed in terms of the Kuhn segment length, a value of l_k=4.2 nm could be established for both solvent conditions [54]. Having thus available a constant value for l_k for inert salt levels of [NaCl]≥0.1 M, we were able to apply Eq. (5) to our experimental data. The dependence of the coil size on the NaCl concentration yielded $f_\alpha = 0.17$ and $w_0 = -0.04$ as fit parameters. Although subjected to a larger uncertainty, values of f_α and w_0 extracted from the molar mass dependent data in 0.1 M NaCl and 1.5 M NaCl agreed with the salt dependent data. It is also interesting to note that the values are compatible with the findings on the PVP chains.

2.4
Effective Charge Density at Low and High Salt Concentrations

2.4.1
Effective Charge Density Determined by the Osmotic Coefficient

The osmotic pressure of a solution of non-interacting molecules or ions at the zero concentration limit is determined by the total molar concentration of solute species. In this hypothetical state, a solution of a polyelectrolyte may contain highly charged polyions at concentration c_p^* and monovalent counterions at a concentration zc_p^*, where c_p^* is the molar concentration of the polyelectrolyte and z is the number of counterions per polyion. Since the total molar concentration is $c_p^*(z+1) \approx c_p^* z$ for $z \gg 1$, the ideal osmotic pressure is

$$\Pi_{id} = RTzc_p^* \tag{19}$$

The product zc_p^* equals Qc_p^m where Q is the fraction of ionizable monomers in the polyelectrolyte and c_p^m is the monomolecular concentration of the polyelectrolyte. To account for nonideality due to electrostatic interaction etc., the osmotic coefficient Φ is introduced and defined as $\Phi = \Pi/\Pi_{id}$ with

$$\Pi = \Phi RTzc_p^* \tag{20}$$

Comparison of equations 19 and 20 shows that we can consider the product $z\Phi$ as an effective charge number z_{eff}, hence $\Phi = z_{eff}/z$. The osmotic coefficient is therefore a measure of the extent to which the dissociated counterions are not independent osmotic entities due to Coulombic interactions [61]. The effective charge density f_{os} is then derived as

$$f_{os} = \Phi Q \tag{21}$$

Manning derived the following expression for Φ based on counterion condensation for infinite dilution

$$\Phi_M = \frac{1}{2\xi} \tag{22}$$

Note that Eq. (22) is the limiting law derived for infinitely long, rod-like polyions.

An alternative theoretical approach is the application of the Poisson-Boltzmann equation on the so-called cell model, assuming a parallel and equally spaced packing of rod-like polyions [62, 63]. This allows one to calculate Φ at finite concentration according to:

$$\Phi_{PB} = \frac{1-\gamma^2}{2\xi} \tag{23}$$

where γ is an integration constant which depends weakly on cell size, viz. concentration. At zero concentration, $\gamma=0$ and $\Phi_{PB}=\Phi_M$. At finite concentration, γ is an imaginary number rising with concentration, so that $\Phi_{PB}>\Phi_M$.

Φ can be measured directly by membrane or vapor pressure osmometry. The application of an alternative method was described recently [64, 65]. It is based on an analysis of the sedimentation equilibrium in an analytical ultracentrifuge, where the solution contains the polyelectrolyte as well as a small concentration of an inert salt. In sedimentation equilibrium, the concentration gradients of both components are coupled via a Donnan-type equilibrium, which is governed by the effective charge number z_{eff} of the polyion. Both concentration gradients can be determined in one experiment, when the polyion and the coion of the salt have sufficiently separated absorption bands in the UV or visible range.

Note that this treatment inherently takes into account the effect of the Donnan equilibrium. The osmotic coefficient obtained therefore is that of the polyelectrolyte with no further Donnan correction term being necessary.

The polyelectrolyte systems studied were sodium polystyrene sulfonate (NaPSS) (with sodium nitroanilinsulfonate being the inert salt) and some of the quaternized polyvinylpyridines listed in Table 2 (with Basic Blue 1=2-chloro-4′,4″-bis-dimethylamino-tritylium chloride being the inert salt). The PVP was quaternized to an extent of 60%, i.e. the reduced charge density is $\xi=1.7$, while the PSS was sulfonated to an extent of 80%, hence $\xi=2.3$.

The point to note is that the polyelectrolytes have a narrow molecular weight distribution with $M_w/M_n<1.2$; this enabled an investigation of the molar mass dependence.

Figure 7 shows the osmotic coefficient of Et-PVP1000X having the same contour length $L_w=1000$ nm but differing with regard to the counterion X, over a limited range of concentrations of the order of $c_p^m \approx 10^{-4}$ M. The data coincide closely ruling out any ion-specific effect. They are markedly lower than the Manning limit ($\Phi_M=0.29$), and they are fairly independent of concentration although a slight increase with rising concentration seems to occur at the higher concentration end.

The osmotic coefficients around 0.15–0.17 correspond to an effective degree of ionization $f_{os}\approx0.1$, which is markedly lower than the value for f_α determined from a fit of the radius of gyration (see Tables 3 and 4), but compare well to the data taken at high salt concentrations as discussed in Sect. 2.4.2. Corresponding data for a NaPSS of contour length around

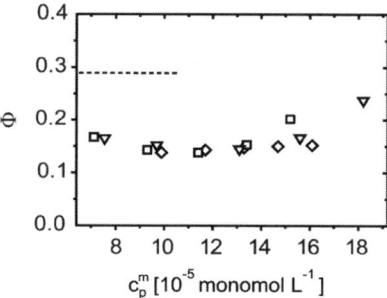

Fig. 7 Concentration dependence of the osmotic coefficient Φ of quaternized PVP, Q=60%. Et-PVP1000I: (◇), Et-PVP1000Br: (□), Et-PVP1000Cl: (▽); the *dashed line* indicates the Manning limit

500 nm (M_w=350.000 g/mol) are shown in Fig. 8. Note that the concentrations are at least ten times larger than for the PVPs. A perceptible, moderate rise of Φ with increasing concentration is observed. This is in qualitative accord with Poisson-Boltzman calculations and also with other experimental data [66].

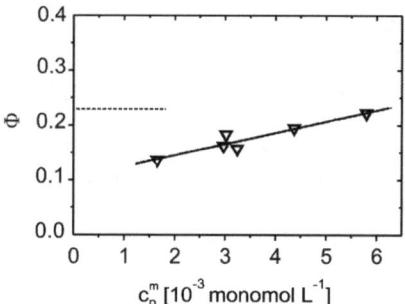

Fig. 8 Concentration dependence of the osmotic coefficient Φ of NaPSS of molar mass 350,000 g/mol, Q=80%; the *dashed line* indicates the Manning limit

Figure 9 shows the dependence of Φ on contour length. The respective concentrations are 10^{-4} M for PVP and ~2 10^{-3} M for PSS. Both polyelectrolytes exhibit a distinct decrease of the osmotic coefficient with rising chain length. At L_w≈100 nm, Φ≈0.25, while for L_w≈1000 nm Φ≈0.15 and keeps going down with further rising contour length. When comparing the PSS and PVP data, it is important to note that not only the concentration regimes (because of experimental accessibility) and contour lengths are different, but also the charge densities. The fact that the data for the two different polyelectrolytes coincide so closely is probably artificial and can result from the cancellation of two effects: the chemical degree of ionization of the

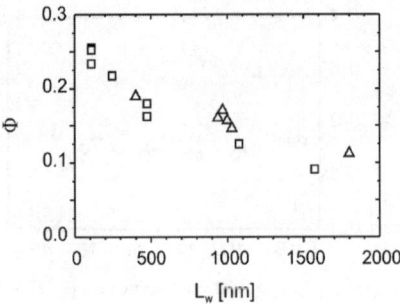

Fig. 9 Dependence of the osmotic coefficient on contour length; NaPSS at $c_p^* \approx 10^{-3}$ mol L^{-1}: ($\diamond$); Et-PVP-X at $c_p^* \approx 10^{-4}$ mol L^{-1}: ($\triangle$)

PVP is lower than that of NaPSS, but the data are taken at considerably lower concentration.

2.4.2
Conductivity Measurements Across the Phase Boundary

In close vicinity to the "salting out" phase transition, the effective charge density was determined by conductivity measurements in combination with the known and estimated electrophoretic mobilities of the counterions and the polyions, respectively. The procedure is based on the fact that, shortly before the precipitation of the polyion, the conductivity comprises contributions of polyions, free counterions and added salt. After precipitation of the polyion by reducing the temperature the conductivity is given solely by the supernatant aqueous salt solution, thus the difference $\Delta\sigma$ is due to polyions and free counterions

$$\Delta\sigma = \sigma_p + \sigma_c \tag{24}$$

For most samples investigated, the temperature interval during which precipitation occurs, is quite small, i.e. 2–5 °C, as shown in Fig. 10. The upper dotted curve in Fig. 10 represents the temperature dependence which would be expected for the polyion solution if no precipitation occurs and if the fraction of free counterions would not increase with temperature, i.e. the increase in conductivity is solely caused by the decrease of the viscosity of water. $\Delta\sigma$ may be utilized to determine the effective charge density f_σ by the equation

$$f_\sigma = \frac{\Delta\sigma}{ec_p^m(\mu_p + \mu_c)} \tag{25}$$

with c_p^m the polyion concentration in monomole/L and μ_p and μ_c the respective mobilities of the polyion and counterion, respectively. Since the mobilities are tabulated and measured at 25 °C also $\Delta\sigma$ is determined at $T = 25$ °C

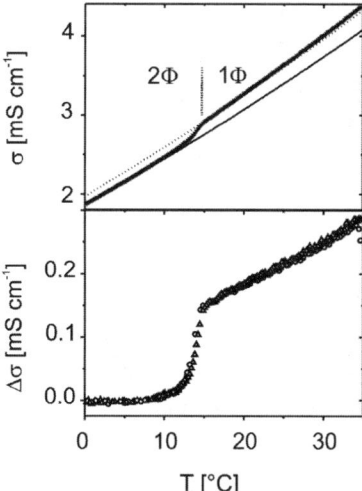

Fig. 10 Temperature dependent conductivity σ for Me-PVP1940I, c_p=5 gL^{-1}, Q=80%. **a** Raw data ($\triangledown$) with salt reference measurements (*solid line*: 0.03 M NaI). **b** Polyelectrolyte conductivity $\Delta\sigma$ after subtraction of the reference measurement. ($\triangle$, $\bigcirc$) repeated measurements

as the difference between the experimental values of the aqueous salt solution and the "hypothetical" upper full curve.

The present samples salt out in NaI solutions, only. For I$^-$ the mobility is μ_I=8 10^{-4} cm^{2}V^{-1} s^{-1}. The polyion mobilities were measured by electrophoretic light scattering, shown in Fig. 11. Although at low salt μ_p linearizes with $c_s^{1/2}$, at high salt concentrations the experimental error becomes larger and the mobility could not be determined close to the phase transition. To this end μ_p is considered to be smaller than 2 10^{-4} cm^{2}V^{-1} s^{-1} and the calculations are performed with μ_p=1±0.8 10^{-4} cm^{2}V^{-1}s^{-1}.

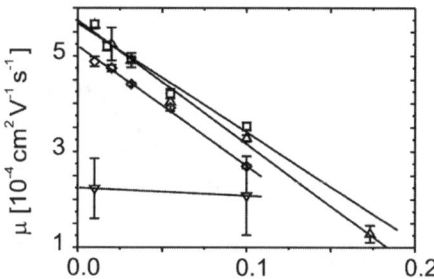

Fig. 11 Electrophoretic mobility of Et-PVP400I at Q=90% ($\square$), Me-PVP1940I at Q=80% ($\triangle$), Me-PVP1940I at Q=50% ($\diamond$), and Bu-PVP1940I at Q=70% ($\triangledown$) as function of added salt concentration

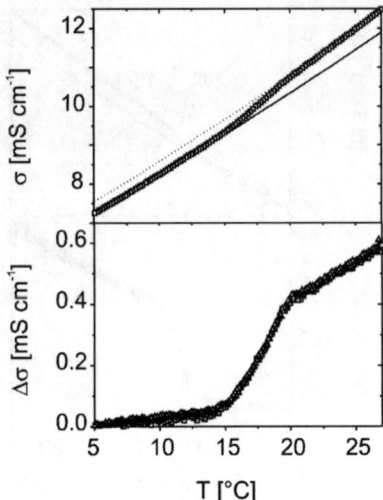

Fig. 12 Temperature dependent conductivity measurements of Et-PVP400I, c_p=12 g L^{-1}, Q=90%. **a** Raw data (□) with salt reference measurements (*solid line*: 0.1 M NaI). **b** Polyelectroyte conductivity after subtraction of the reference measurement. (□, △) repeated measurements

Fortunately, this large uncertainty in μ_p does not influence the value of f_σ significantly because $\Delta\sigma$ almost exclusively originates from the I$^-$ mobility. Finally it should be mentioned that not all investigated polyion systems precipitate within a small temperature interval. Figure 12 shows that the conductivity of sample Et-PVP1940I exhibits an initial sharp drop of σ followed by a pronounced tailing towards lower temperatures. In such case the conductivity drop was determined by fitting the conductivity of the two phase solution by the reference conductivity of salt solution below the precipitation temperature. In some rare cases observed particularly at low chemical charge densities no sharp drop of σ is observed but rather a continuous

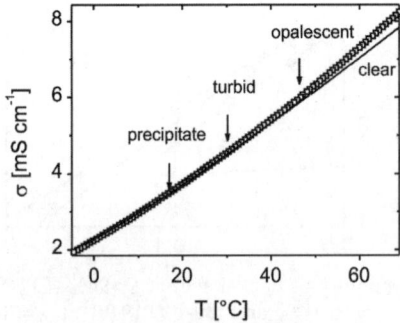

Fig. 13 Temperature dependent conductivity measurements for Et-PVP1940I at Q=40% (□) with a salt reference measurement (*line*)

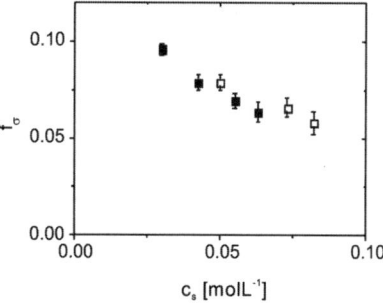

Fig. 14 Effective charge density f_σ close to the precipitation boundary as a function of NaI concentration; (■): MePVP1940I, Q=80% and (□): BuPVP1940I, Q=70%

transition as shown in Fig. 13. For these samples f_σ could not be determined. The origin of this continuous and slow precipitation process could not yet be identified. Phenomenologically, colloidal particles form in a first step, which slowly aggregate and eventually precipitate.

Despite the strong scatter of the data there is a tendency of f_σ to decrease with increasing c_s as graphically shown in Fig. 14.

2.4.3
Discussion of the Effective Charge Densities

The effective charge densities derived from the osmotic coefficient f_{os} (see Sect. 2.4.1) is much lower than expected from Manning condensation or by the Poisson-Boltzmann theory. It compares well with the low effective charge densities determined by the conductivity measurements at high added salt concentrations, i.e. at the "salting out" phase transition, which lie in the range of $0.1 < f_\sigma < 0.05$. This seems to indicate that the effective charge density would not vary much with added salt concentration, although the conductivity measurements seem to show a decrease of f_σ with increasing salt concentration. The low charge densities discussed seem to be in contradiction to the much higher charge densities f_α derived from the polyion dimensions utilizing Muthukumars excluded volume expression given by Eq. (5). At this point the question ought to be addressed whether the three methods applied for the determination of the effective charge density could be expected to yield similar results. Muthukumars expression calculates the expansion of the polyion dimensions due to the electrostatic excluded volume. Thus, f_α represents the charge density which is experienced by "probe charges" located in the close vicinity of the polyion chain. This charge density f_α could be significantly larger than f_{os} derived from the osmotically active counterions which predominantly consist of the free counterions far away from the chain backbone. Finally, the evaluation of f_σ requires the assumption that the counterion mobility is not affected by the presence of the polyion. Electrostatic coupling to the polyion dynamics may slow down the counterions slightly.

In summary the discussion above appears to provide a crude and qualitative explanation of the experimental charge densities. For more detailed recent progress by computer simulations in this field the reader is referred to theoretical contributions in this volume. However, the determination of the effective charge densities remains a highly delicate issue and the discussion above may at most provide some qualitative trends which, however, are based on several assumptions.

3
Aspects of Phase Behavior of Quaternized Polyvinylpyridine and Sodium Polyacrylate

3.1
Salting Out and Salting in of Poly(2-vinylpyridinium) Cations

As already indicated in Sect. 2.2.2 the salting out concentration depends on the chemical nature of the counterions.

For NaI a peculiar phenomenon is observed. After the polyelectrolyte precipitates at a certain concentration of added NaI ("salting out") it dissolves again upon further increase of the salt concentration ("salting in"). For the polyvinylpyridinium system this phenomenon is only observed for added salt containing iodide, whereas long ago the salting in was also observed for anionic polyelectrolytes [21, 22].

The location of the phase boundaries as function of added salt was monitored as a function of temperature, charge density, chain length and hydrophobicity of the polyelectrolyte. The polyion concentration was kept constant at 0.1%. However, the results did not change significantly if c_p was increased to 2%. Higher polyion concentrations were not investigated systematically.

It should be noted, that the miscibility gaps discussed throughout the paper are not to be confused with classical coexistence curves, i.e., the miscibility gap does *not* refer to the phase separation into a salt-rich and a salt-poor phase, but rather to the solubility/precipitation of the polyelectrolyte at constant c_p=0.1%.

The complete determination of the solubility diagram is prohibited by two limitations: the accessible temperature range in water is naturally limited from 0 °C to 100 °C and the upper salt concentration is limited by the solubility of NaI in water.

The **chemical charge density** is the most important parameter governing the size of the region where the polyion precipitates (two phase region). The salting in boundary is only observed for high degrees of alkylation. Above a chemical charge density of about 70% the phase boundaries almost coincide for samples with identical alkyl substituents. Figure 15 shows this for Et-PVP1940I. The miscibility gap reduces with increasing charge density

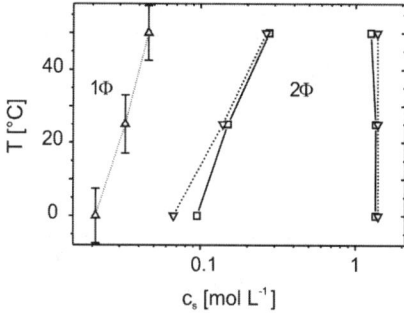

Fig. 15 Phase boundaries of Et-PVP1940I in aqueous NaI solution with different chemical charge densities Q=40% ($\triangle$), 70% ($\triangledown$) and 80% ($\square$)

(Fig. 16) even above the Manning limit, corresponding to a charge density of 0.35.

For highly quaternized polyelectrolytes, temperature dependent conductivity measurements exhibit a sudden drop of conductivity while crossing

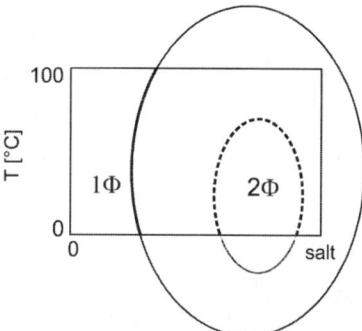

Fig. 16 Sketch of the hypothetic variation of the miscibility gap with increasing chemical charge density (experimental window indicated by *rectangle*)

the phase boundary. In contrast, polyelectrolytes with smaller chemical charge density precipitate continuously over a large temperature interval.

The **hydrophobicity** also influences the size of the two phase region, but not as pronounced as the charge density. Figure 17 shows this by comparison of Bu-PVP1940I and Et-PVP1940I for similar chemical charge density. The more hydrophobic Bu-PVP is salted out at about half the salt concentration needed for Et-PVP.

The introduction of ether groups into the side chains was used for further increase of hydrophilicity. Scheme 2 shows the hydrophilic quaternization reagent methoxy-ethoxy-butyl-tosylate (ME1B-Tos). The corresponding polyelectrolyte ME1B-PVP shows an upper critical solution temperature

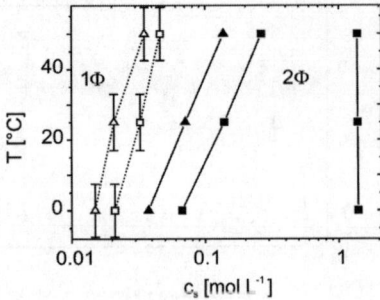

Fig. 17 Phase boundaries of Et-PVP1940I (■, □) and Bu-PVP1940I (▲, △) in aqueous NaI for different degrees of quaternization ($Q=70\%$ solid symbols and 40% open symbols)

(Fig. 18), exhibiting a behavior similar to PEO [67, 68], which is governed by hydrogen bridging between water and monomer units [69]. Hence, further increase of hydrophilicity moves the miscibility gap up, as shown in Fig. 19.

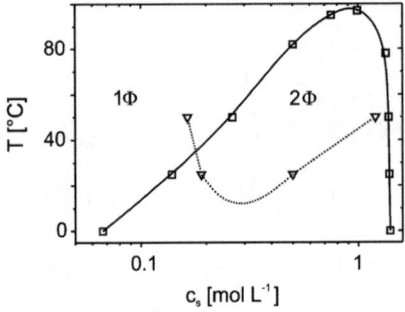

Fig. 18 Upper and lower critical solution temperature of ME1B-PVP1940Tos, $Q=70\%$ (▽) and Et-PVP1940I, $Q=70\%$ (□) respectively.

Scheme 2 Structure of the hydrophilic quaternized agent methoxy-ethoxy-butyl-tosylate ME1BPVP1940Tos

The **contour length** of the polyelectrolyte shows only a minor influence on the miscibility gap. With decreasing size of the chain length the two phase region increases and is moved to slightly lower salt concentrations for intermediate degrees of quaternization. Moreover the temperature dependence of the phase boundary decreases.

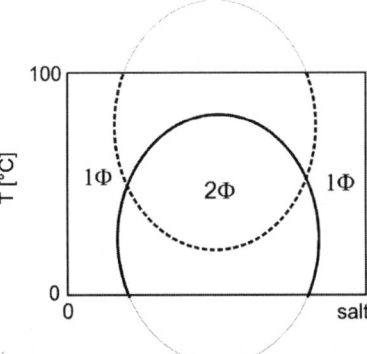

Fig. 19 Sketch of the hypothetic shift of the miscibility gap for hydrophilic substituents (experimental window indicated by *rectangle*)

For high chemical charge densities the miscibility gap coincides independently of the contour length (Fig. 20), i.e. Et-PVP400I exhibits the same behavior as Et-PVP1940I even though the contour length is varied by a factor of 5.

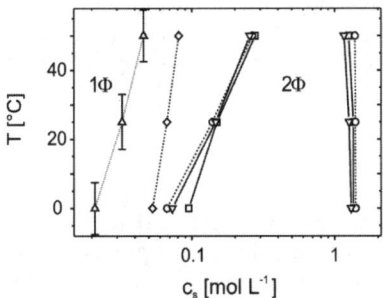

Fig. 20 Phase boundaries of Et-PVP in aqueous NaI under variation of chemical charge density and contour length Et-PVP400I and Q=90% (○), Et-PVP1000I and Q=60% (◇). (Other *symbols* as in Fig. 15)

Finally, the most likely mechanism for the "salting in" effect ought to be discussed. A simple explanation would be that NaI represents "hypothetically" a good solvent for the polyion. Thus with increasing volume fraction of NaI the polyion becomes more soluble which would be equivalent to an increase w_0, due to an increase of the solvent quality. A more likely mechanism of the salting in seems to be a charge reversal i.e. the formation of a charge transfer complex between the pyridinium and iodide ion. This is long known for low molecular weight compounds [70,71], but in the case of quaternized PVP additional substituents causing a bathochromic shift would be needed for detection of the charge transfer complex via UV absorption.

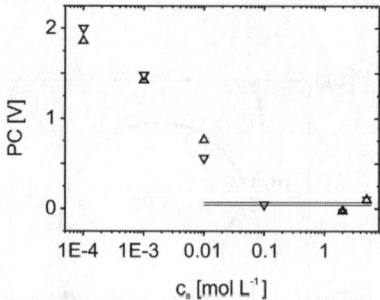

Fig. 21 Particle charge of various X-PVP against salt concentration, errors within size of the symbol as indicated in the high salt regime. *Lines* indicate different salt reference measurements. Me-PVP1000Br, $Q=60\%$ ($\triangledown$) and Me-PVP1940I, $Q=70\%$ ($\triangle$)

The salting in regime could only be investigated via a particle charge detector. This easy set up induces an electrical field by an oscillating flow. The electrical field relates to the particle charge, but the method cannot give an absolute value [72]. In Fig. 21 is shown, that close to the phase boundaries of the system, the polyelectrolyte exhibits almost no charge and becomes slightly negatively charged above the "salting in" regime. However, this data scatters around zero, which prohibits a final proof of charge reversal. This point needs further investigation. It should be noted, that the concentrated iodide solution is highly sensitive to oxidation i.e. by light or electrical fields which prohibits electrophoretical experiments within the salting in regime.

3.2
L-type Precipitation of Sodium Polyacrylate

Wall and Drenan [24] were among the first to notice the impact of earth alkaline cations on the solution behavior of polyacrylic acid. They added solutions of $M(OH)_2$ with M=Ca, Sr, Ba to solutions of polyacrylic acid in distilled water and observed precipitation of MPA. Inspired by the amorphous appearance of the precipitate, they used the term gelation. The smaller the acid concentration was, the more M per COO^- group was required to induce precipitation. According to Wall and Drenan, monodentate binding of M^{2+} is necessary to cause bridging to a COO^- group from a different polymer chain. The likelihood for such intermolecular bidentate complexes increases with increasing PA concentration. At a distinct acid concentration, gelation efficiency increased according to Ca<Sr<Ba, indicating that gelation is not only a matter of electrostatic interactions.

Following this paper, two pioneering publications revealed the major aspects of this precipitation [23,25]. In both investigations, aqueous solutions of the sodium salt of poly-carboxylic acids were used instead of the polyacid. Michaeli [25] investigated in a highly systematic way the influence of an inert salt NaCl and the degree of ionization Q of the polyelectrolyte chains on the precipitation behavior of sodium polymethacrylic acid (NaPMA). He was able to show that the amount of Ca^{2+} required to precipitate NaPMA increased linearly with increasing NaPMA according to

$$[M^{2+}]_c = m + r_0[COO^-]_c \qquad (26)$$

In Eq. (26), $[M^{2+}]_c$ and $[COO^-]_c$ denote molar concentrations of M^{2+} and monomer respectively required for precipitation. The intersection of the line and its slope were best described according to

$$m = \frac{0.0034}{Q^2} + 0.13[NaCl] \qquad (27a)$$

$$r_0 = Q(0.8 - 0.7[NaCl]) \qquad (27b)$$

For fully dissociated NaPMA, the straight lines approached the origin. Addition of NaCl caused m to increase proportional to $[NaCl]$. At the same time the slope r_0 gradually decreased. The latter effect was compatible with the recent finding [73] that the precipitates included a fraction of monovalent cations which obviously increase with increasing $[NaCl]$. If the intersection is extrapolated to $[NaCl]=0$, a quasi solubility product $mQ^2=0.0034$ can be inferred from Eq. (27a). This law of mass action, however, refers to the separated coils as the "reaction containers" for the precipitation process. The underlying reaction $M^{2+} + 2COO^- \rightarrow M(COO)_2$ changes the solubility of the PMA chains. According to Michaeli [25] it is this solubility change which leads to the precipitation rather than an intermolecular cross linking.

Ikegami and Imai [23] extended our knowledge on NaPA, confirming Eq. (26) for $Q=1$ and $[NaCl]=0$ and $M=Ag^+$, Mg^{2+}, Ca^{2+}, Ba^{2+} and La^{3+}. If self-ionization was completely depressed by adding HCl, precipitation of PA did not follow Eq. (26) any more but required much larger amounts of bivalent metal cations (H-type), independent of $[COO^-]$. At an intermediate neutralization of $Q=0.25$, Ba^{2+} resulted still in an L-type precipitation whereas Mg^{2+} could only induce a H-type precipitation [23].

Inspired by the work of Michaeli [25] and Ikegami and Imai [23], we decided to investigate the coil conformation of NaPA chains when approaching the precipitation limit [74-76]. First investigations were performed with Ca^{2+} as a bivalent metal cation [74, 75]. The approach to the phase boundary was achieved by either gradually decreasing the NaPA concentration at a constant Ca^{2+} concentration (route 1) or by increasing the Ca^{2+} concentration at a constant NaPA concentration (route 2) [75, 76]. Along both routes, series of intermediates could be investigated with combined static and dynamic LS. The main feature of any series is the varying ratio of $[Ca^{2+}]/[COO^-]$. Within a series, the overall concentration of positive charges stemming from the inert salt and Ca^{2+} was kept constant. The bivalent cations were introduced by replacing the corresponding amount of Na^+ ions. The extent of replacement did not exceed 20% and in most cases it was kept below 10%.

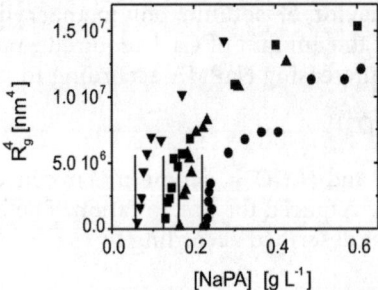

Fig. 22 Static light scattering close to the precipitation threshold. Measurements are performed at constant $[Ca^{2+}]$ by decreasing $[NaPA]$ (route 1). The inert salt level is 0.1 M NaCl. Squared mean square radius of gyration is plotted vs. $[NaPA]$. The *symbols* denote four different Ca^{2+} concentrations: 3.25 mM (▼); 3.5 mM (■); 3.75 mM (▲); 4.0 mM (●). *Vertical lines* indicate precipitation limits which appear as *open circles* in Fig. 23

Replacement of cations at a fixed concentration of positive charges kept the effect of regular screening approximately constant and the impact of bivalent metal cations was isolated [54,74,75]. As an example, Fig. 22 outlines the results from static LS for four series of measurements with a single NaPA sample. Data are represented in terms of the initial slopes of the scattering curves, which are proportional to the mean squared radius of gyration R_g^2. Each series approaches the phase boundary at a different concentration of Ca^{2+} along route 1. Within each series, R_g^2 decreases with decreasing $[COO^-]$. In order to amplify the effect, the square of R_g^2 is plotted indicating a threshold concentration $[COO^-]_c$ where R_g^2 falls short of the experimentally accessible limit. If this concentration threshold was crossed, the apparent molar mass started to significantly increase and a second mode appeared in dy-

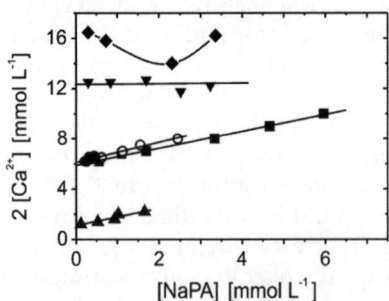

Fig. 23 Phase boundaries for PA2 at four different NaCl levels: 0.01 M (▲) 0.1 M (■) 0.8 M (♦) and 1.5 M (▼). An additional phase boundary is included for sample PA1 in 0.1 M NaCl (●, ○). The *open circles* indicate the respective situation for the measurement series discussed in detail in Fig. 22

Table 5 Parameters of Eq. 26 for phase boundaries of the system NaPA with Ca^{2+} at $T=25$ °C

[NaCl]	Sample	$M_w[10^6 \text{ gmol}^{-1}]^a$	$m[\text{mmolL}^{-1}]$	r_0
1.5	PA2	3.3	6.13	0
0.1	PA2	3.3	3.05	0.405
0.1	PA1	0.95	2.94	0.34
0.01	PA2	3.3	0.549	0.345

[a] Molar mass values are based on $dn/dc=0.167$ cm^3/g

namic LS. Both effects indicated aggregation and served as criteria to exclude further evaluation of the respective data in terms of single chain behavior. In fact, the respective solution gradually became turbid [74, 75].

The phase boundaries thus recorded for Ca^{2+} are summarized in Fig. 23. Parameters from Eq. (26) are represented in Table 5. The results for two more NaPA samples in 0.1 M NaCl from an earlier publication [74] (m=3.12 mmol/L, 3.4 mml/L and r_0=0.27 and 0.28 respectively) agree fairly well with the data in Table 5. The only value which seems to deviate slightly is r_0=0.41 in Table 5, specially if we bear in mind that r_o is expected to decrease with increasing [NaCl]. The present data are also compatible with recent findings by Pochard et al. [73] (r_0=0.33) and by Sabbagh and Delsanti [77] (r_0=0.36), both in distilled water. However, Eq. (26) fails to describe the precipitation behavior of NaPA if the inert salt level significantly exceeds 0.1 M NaCl. The phase boundary even exhibits a minimum if [NaCl]=0.8 M. It is also worth being mentioned in addition, that the extension of the one phase regime limited by the curve $[M^{2+}]$ versus $[COO^-]_c$ is largest in the regime around [NaCl]=0.8 M and shrinks on further addition of NaCl.

In this context, reference is given once more to the results by Sabbagh and Delsanti [77] on solutions free of inert salt. They could show that deviations from Eq. (26) also occurred if the polymer concentration was low enough. A minimum in the diagram $[M^{2+}]$ versus $[COO^-]_c$ appeared which was shifted to larger values of $[COO^-]_c$, the smaller the molar mass of the NaPA samples became.

From the experiments just outlined [74–76], a few points are worth being emphasized: A powerful procedure was developed to gradually approach phase boundaries of polyelectrolyte precipitation. The approaches can be performed in a highly systematic manner and lead to states which are located extremely close to the precipitation threshold. Approaches could successfully be accompanied by LS experiments. The experiments demonstrated, that the polyelectrolyte chains shrank dramatically in size immediately before the phase boundaries were reached. A sudden increase of the scattering intensity indicated the phase boundaries. These developments give rise to the hope that intermediates may be revealed which have not become accessible in preceding investigations [78–81].

4
Size and Shape of Collapsing Polyelectrolytes by Light Scattering

4.1
Ca^{2+} Induced L-Type Precipitation of Sodium Polyacrylate

As described in Sect. 3.2, the size of the NaPA coils shrink drastically when the precipitation line is approached. This is revealed by a decrease of the radius of gyration R_g and the hydrodynamically effective radius R_h. In exploiting this effect, data from various approaches to phase boundaries belonging to different inert salt levels and NaPA samples had been collected. Facing such an extensive set of data, a meaningful tool for its interpretation seems highly desirable.

The two dimensionless parameters α and ρ appear to be especially helpful. α is the ratio of the radius of gyration of any intermediate along the approach to the phase boundary and the radius of gyration of the unperturbed chains, i.e., the Θ-state (see Eq. (4). The latter were determined experimentally in 1.5 M NaCl. The second parameter is the shape sensitive ratio ρ of the radius of gyration and the hydrodynamic radius (see Eq. (17).

A plot of the two parameters as ρ versus α promises new insight in the M^{2+} induced shrinking mechanism. As long as the mechanism is the same, a single curve could be expected, independent of the NaPA sample, of the inert salt level, of the type of the bivalent cation and of the special location where the phase boundary would be approached.

Figure 24 summarizes plots of ρ versus α at three different levels of NaCl. The NaCl concentrations are 1.5 M, 0.1 M and 0.01 M respectively. Whereas Figs. 24a and 24c exhibit data from a single NaPA sample (M_w=3.3 10^6 g/mol), Fig. 24b even includes data from two different samples (M_w=0.95 10^6 g/mol and M_w=3.3 10^6 g/mol). At the inert salt level of 1.5 M NaCl, the shrinking of NaPA chains due to the addition of Ca^{2+} began at a state already as small as its unperturbed dimensions. Contrary to this, the NaPA coils at the two lower salt levels of 0.1 M and 0.01 M NaCl are highly extended in the absence of Ca^{2+} leading to much larger shrinking ratios if referred to the M^{2+}-free state. However, only data from intermediates where the shrinking had reached an extent which made $R_g<R_{g,\Theta}$ have been used for the plots. In all three Figures an additional curve is included. This curve corresponds to a plot inferred from experimental data of poly-(N-isopropylacryamide) (PNIPAM) in water [82]. At T=30.6 °C, water is a Θ-solvent for PNIPAM and the resulting curve represents the collapse of a neutral polymer chain if the Θ-point is crossed. The curve begins with ρ=1.5 at α=1, gradually decreasing to a value of ρ=0.77 at shrinking ratios close to α=0.2, which is compatible with a coil to sphere transition. The curve thus serves as an excellent reference curve.

At $[NaCl]$=1.5 M, the phase boundary is a horizontal line and the approach had to be performed along route 2. Unfortunately, the results turned out not to be very instructive because the experimentally accessible shrink-

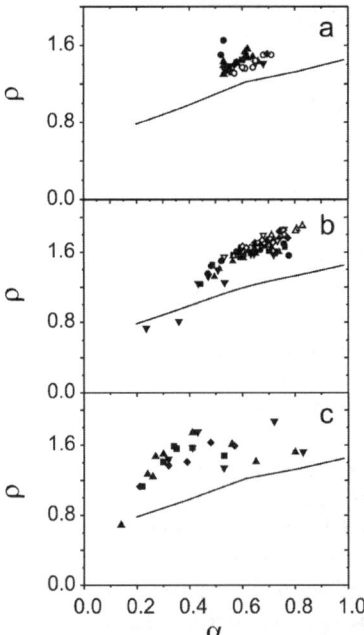

Fig. 24 ρ parameter vs. shrinkage ratio α at three different NaCl concentrations [75]: 1.5 M (*a*) 0.1 M (*b*) and 0.01 M (*c*). Experiments with sample PA2 are denoted as **a** c_p= 3.234 mmol L^{-1} (■); c_p=0.29106 mmol L^{-1} (●); c_p= 0.84043 mmol L^{-1} (▲); c_p=3.3936 mmol L^{-1} (▼); c_p=2.516 mmol L^{-1} (○); c_p=1.6968 mmol L^{-1} (◆); **b** c_p=3.25 mmol L^{-1} (■); c_p=3.5 mmol L^{-1} (●); c_p=3.75 mmol L^{-1} (▲); c_p=4 mmol L^{-1} (▼); **c** c_s=0.6 mmol^{-1} (▲); c_s=0.8 mmol^{-1} (▼); c_p=1 mmol L^{-1} (◆); c_s=0.7 mmol^{-1} (■). PA1 was only investigated in the presence of 0.1 M NaCl-content at four different Ca^{2+} contents: c_s=3.5 mmol^{-1} (◆); c_s=4 mmol^{-1} (∇); c_s=4.5 mmol^{-1} (△); c_s=5 mmol^{-1} (◇). The limit of a sphere like shape is reached in the 0.1 M NaCl and 0.01 M NaCl solutions. The *curve* represents literature data for a coil-globule-transition of neutral polymers below Θ-temperature [82]

ing regime was too narrow (Fig. 24a). If the extent of shrinking dropped below 0.5, the samples usually started to aggregate, preventing further measurement of single chain behavior. The situation became much more interesting however, at an inert salt level lower by more than an order of magnitude.

Figure 24b represents the shrinking at [*NaCl*]=0.1 M. This time, most of the measurement series were performed along route 1. Independent of the NaPA sample and the Ca^{2+} concentration, the data followed the same trend. They started at $\rho\sim 1.6$ as expected for unperturbed NaPA chains (Chapter 2.3), being slightly above ρ-ratios corresponding to the Θ-state of monodisperse neutral coils. In a few rare cases they were getting close to 0.77 at $\alpha\sim 0.3$. Clearly, the NaPA chains adopt the shape of a sphere, prior to precipitation with Ca^{2+}. The difference between the present trend and the reference curve based on PNIPAM may be due partly to the larger polydispersity of the NaPA samples.

Finally, in Fig. 24c the inert salt level was decreased by another order of magnitude. Although scatter of the results was somewhat larger than at $[NaCl]=0.1$ M, data at $[NaCl]=0.01$ M again followed a unique trend, independent of the Ca^{2+} concentration selected for the measurement series along route 1. As in 0.1 M NaCl, the ρ-values started at $\rho \sim 1.6$ for $\alpha=1$ and reached the sphere limit at the largest extent of shrinking achieved by experiment. However, it is most striking that the trend in 0.01 M NaCl was different from the one observed at 0.1 M NaCl. Whereas in 0.1 M NaCl the sphere limit was reached at $\alpha \sim 0.3$, it was approached only at $\alpha \sim 0.1$ in 0.01 M NaCl. Clearly, the experimentally accessible extent of shrinking was much larger for the lower concentration of inert salt. At the same time the ρ-values kept constant within experimental error over a large shrinking regime of $0.3 < \alpha < 1.0$.

This effect is compatible with at least two alternative mechanistic models at 0.01 M NaCl: (i) the shrinking process passes self similar intermediates over a wide range of α; (ii) the mechanism comprises two opposing effects, a condensation of monomers decreases ρ and an increasing anisotropy of the collapsing species increases ρ. No matter which of the proposed models describes the actual process, the mechanism seems to be different from the one taking place at the higher inert salt level. This conclusion can be drawn unambiguously from the different trends shown in Figs. 24b and 24c. In spite of mechanistic differences, the limiting structures are equal in both inert salt levels.

After having discussed the influence of the inert salt level on the Ca^{2+} induced coil collapse, we would like to draw attention to another interesting aspect. As found by Michaeli [25] and confirmed by our own data [75,76], addition of inert salt increases the solubility of CaPA. This immediately poses the question of how the addition of inert salt affects the size and shape of a collapsed NaPA chain.

To answer this question, NaPA chains were collapsed by Ca^{2+} ions along route 1 at two different phase boundaries. The collapsed state was established by ρ-values significantly smaller than 1. Having approached this state, the NaCl concentration of the sample was increased at constant concentrations of all other components. This procedure was performed with three different series starting from three different collapsed samples. Two belonged to the same phase boundary at 0.01 M NaCl differing only in the values for $[Ca^{2+}]_c$ and $[COO]_c$. The results were striking.

In all three experimental series, addition of NaCl caused a gradual increase of the coil dimensions which approached or even slightly crossed the unperturbed radii. This can be explained as follows. The Ca^{2+} ions are expected to be bound to the COO^- groups via complex bond formation which render the chain hydrophobic and insoluble in water. Addition of NaCl generates a swamping excess of Na^+ ions which gradually replace the Ca^{2+} cations. Being bound less tightly to the COO^- groups than the Ca^{2+} ions, the

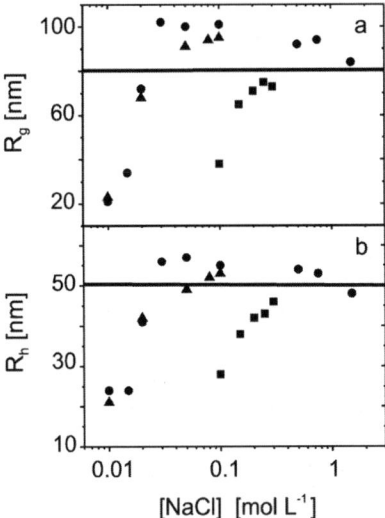

Fig. 25 Radii of gyration R_g (a) and hydrodynamic radii R_h (b) versus NaCl concentration [75] for PA2 at c_s=0.9 mmol L^{-1} and c_p=1.2234 mmol L^{-1} (●); c_s=1.2 mmol L^{-1} and c_p=2.1809 mmol L^{-1} (▲); c_s=4.0 mmol L^{-1} and c_p=2.5 mmol L^{-1} (■). The *lines* indicate unperturbed dimensions of PA2 at 1.5 M NaCl

chains recover part of their hydrophylicity and start to expand. Coil expansion for all three series are summarized in Fig. 25a,b. Transformed into ρ and α data, all three series yield a single trend shown in Fig. 26. This indicates that expansion proceeds according to the same mechanism for all three series of measurements and at the same time proves the validity of the ρ versus α plot!

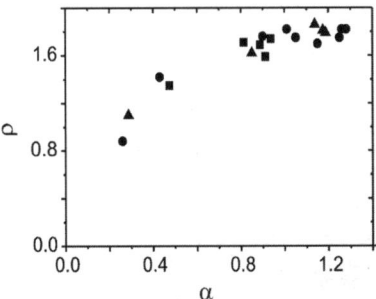

Fig. 26 ρ-parameter vs shrinkage ratio α for the three measurement series [75] of Fig. 25 at c_s=0.9 mmol L^{-1} and c_p=1.22 mmol L^{-1} (●) c_s=1.2 mmol L^{-1} and c_p=2.18 mmol L^{-1} (▲); c_s=4.0 mmol L^{-1} and c_p=2.5 mmol L^{-1} (■)

4.2
Comparison of Ca^{2+}, Sr^{2+} and Ba^{2+} Ions

The impact of the type of earth alkaline cation on the shrinking mechanism is an interesting question by itself. Moreover, knowledge in this field is relevant for SAXS measurements. The earth alkaline cations with higher electron density might reflect the shape of the collapsing chains if they are bound to the very chains [77].

In the present chapter, a preliminary investigation of the influence of homologous cation variation on the structural transformation shall be given. This comparative investigation was performed with Ca^{2+}, Sr^{2+} and Ba^{2+} at an inert salt level of 0.01 M NaCl. In all cases, the phase boundary was approached along route 1 at different concentrations of M^{2+}. All results stem from the same NaPA sample and are summarized in Fig. 27 and Table 6.

Clearly, the phase boundaries lie on top of each other. A closer look reveals a slight decrease in the slope r_0 of Eq. (26) according to $Ca^{2+} > Sr^{2+} > Ba^{2+}$, which indicates that the larger the bivalent earth alkaline cation is, the smaller is the stoichiometric amount of M^{2+} necessary to precipitate NaPA. Although based on a different method, the present results suggest a comparison with data from Pochard et al. [73]. In doing so, we have to keep in mind that our own r_0 values are of fair accuracy at best because they were evaluated from slopes based on a few data points only. Still, this trend in r_o is the opposite to the observation of Pochard et al. [73]. They found a decrease of the amount of M^{2+} per COO^- function at the precipitation threshold, if Ca^{2+}

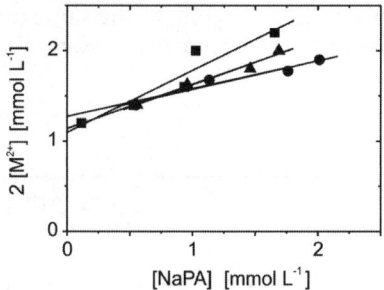

Fig. 27 Phase boundaries for three different earth alkaline cations in 0.01 M NaCl: Ca^{2+} (■); Sr^{2+} (▲); Ba^{2+} (●). The polymer sample is PA2

Table 6 Parameters of Eq. 26 for phase boundaries of the system PA2 with M^{2+} denoting earth alkaline cations at T=25 °C

[NaCl]	M^{2+}	m[mmolL^{-1}]	r_0
0.01	Ca^{2+}	0.549	0.345
0.01	Sr^{2+}	0.563	0.249
0.01	Ba^{2+}	0.622	0.160

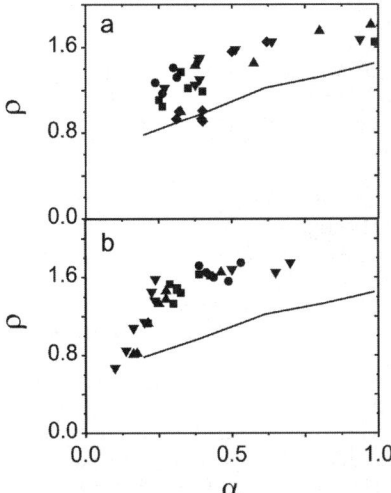

Fig. 28 ρ-parameter vs. shrinkage ratio α at 0.01 M NaCl for two different earth alkaline cations: Sr^{2+} (a) and Ba^{2+} (b). Different approaches to the phase boundary are denoted in **a** c_p=0.957 mmol L^{-1} (■); c_p=0.559 mmol L^{-1} (●); c_s=1.0 mmol L^{-1} (▼); c_p=0.9 mmol L^{-1} (▲); c_p=1.5 mmol L^{-1} of PA1 (◆); **b** c_p=0.55 mmol L^{-1} (▼); c_p=1.13 mmol L^{-1} (◆); c_s=1.76 mmol L^{-1} (●); c_s=0.95 mmol L^{-1} (▼). The *curve* represents literature data for a coil-globule-transition of neutral polymers below Θ-temperature [82]. In both figures specification of c_s or c_p corresponds to route 1 or to route 2 respectively. Unless otherwise mentioned the polymer sample is PA2

is exchanged by Ba^{2+}. Neutralization of the PA chains with M^{2+} increases the hydrophobic nature of the chain and according to Pochard et al. [73], an increase in the hydrophobic nature of the bidentate complex between M^{2+} and COO^- lowers the critical amount of M^{2+} per COO^- necessary to precipitate the MPA salt. As a consequence, their results suggest a hydrophobicity which is larger for Ca^{2+} than for Ba^{2+}.

Turning to the shrinking mechanism, we are faced with data of varying quality. Barium, as the largest cation in the row, exhibited a trend (Fig. 28 a) which was of better quality than Ca^{2+} (Fig. 24c). Yet, for both cations, constant ρ ratios were observed over a regime of shrinking of $0.3<\alpha<1.0$ and the sphere limit was reached only at $\alpha<0.2$. This points to a single mechanism of coil shrinking in 0.01 M NaCl which includes either self similar or increasingly anisotropic intermediates prior to reaching the sphere limit. Unfortunately data from Sr^{2+} cations scattered slightly more than those of Ca^{2+} (Fig. 28b) and do not allow clear assessment of the type of mechanism.

References

1. Yamakawa H (1971) Modern Theory of Polymer Solutions. Harper and Row, New York
2. Fujita H (1990) Polymer Solutions. Elsevier, New York
3. Meewes M, Ricka J, de Silva M, Nyffenegger R, Binkert T (1991) Macromolecules 24:5811 and references therein
4. Wu C, Zhou S (1995) Macromolecules 28:5388
5. Wu C, Zhou S (1995) Macromolecules 28:8381
6. Wang X, Qiu X, Wu C (1998) Macromolecules 31:2972
7. Chu B, Ying Q, Grosberg AY (1995) Macromolecules 28:180
8. Chu B, Ying Q (1996) Macromolecules 29:1824
9. Borochov N, Eisenberg H (1994) Macromolecules 27:1440
10. Förster S, Schmidt M, Antonietti M (1992) J Phys Chem 96:4008
11. Ise N, Okubo T (1980) Acc Chem Res 13:303
12. Ise N, Okubo T, Yamamoto K, Kawai H, Hashimoto T, Fujimura M, Hiragi Y (1980) J Am Chem Soc 102:7901
13. Dosho S, Ise N, Ito K, Iwai S, Kitano H, Matsuoka H, Nakamura H, Okumura H, Ono T, Sogami IS, Ueno Y, Yoshida H, Yoshiyama T (1993) Langmuir 9:394
14. Antonietti M, Briel A, Förster S (1996) J Chem Phys 10:7795
15. Gröhn F, Antonietti M (2000) Macromolecules 33:5938
16. Harada T, Matsuoka H, Ikeda T, Yamaoka H (2000) Langmuir 16:1612
17. Zhang Y, Douglas J F, Ermi BD, Amis EJ (2001) Chem Phys 114:3299
18. Flory PJ, Osterheld JE (1954) J Phys Chem 58:653
19. Eisenberg H, Woodside D (1962) J Chem Phys 36:1844
20. Takahashi A, Kato T, Nagasawa M (1967) J Phys Chem 71:2001
21. Eisenberg H, Mohan GR (1959) J Phys Chem 63:671
22. Eisenberg H, Casassa EF (1960) J Polym Sci 47:29
23. Ikegami A, Imai N (1962) J Polym Sci 56:133
24. Wall FT, Drenan JW (1951) J Polym Sci 7:83
25. Michaeli I (1960) J Polym Sci 48:291
26. Olvera de la Cruz M, Belloni L, Delsanti M, Dalbiez JP, Spalla O, Drifford M (1995) J Chem Phys 103:5781
27. Wittmer J, Johner A, Joanny JF (1995) J Phys (Paris) II 5:635
28. Imai N, Onishi T (1959) J. Chem. Phys. 30:1115
29. Oosawa F (1971) Polyelectrolytes, Marcel Dekker, New York
30. Manning GS (1978) J. Chem. Phys. 51:924, 934, 3249, (1969) Biophys. Chem. 9:65
31. Manning GS (1978) Biophys Chem 9:65
32. Flory PJ (1953) J Chem Phys 21:162
33. Muthukumar M (1987) J Chem Phys 86:7230
34. Muthukumar M (1996) J Chem Phys 105:5183
35. Odijk T (1977) J Polym Sci Polym Phys Ed 15:477
36. Odijk T, Houwaart AC (1978) J Polym Sci Polym Phys Ed 16:627
37. Skolnick J, Fixman M (1977) Macromolecules 10:944
38. Fixman M, Skolnick J (1978) Macromolecules 11:863
39. Le Bret (1982) J Polym Phys 76:6248
40. Barrat JL, Joanny JF (1996) Adv Chem Phys XCIV: 1
41. Li H, Witten TA (1995) Macromolecules 28:5921
42. Ha BY, Thirumalai D (1995) Macromolecules 28:577
43. Beer M, Schmidt M, Muthukumar M (1997) Macromolecules 30:8375
44. Kuhn W, Künzle O, Katchalsky A (1948) Helv Chim Acta 31:1994
45. De Gennes PG, Pincus P, Velasco RM, Borchard F (1976) J Phys France 37:1461
46. Reed WF, Ghosh S, Medjadhi G, Francois J (1991) Macromolecules 24:6189
47. Des Cloizeaux J, Jannink G (1990) Polymers in Solution, Their Modeling and Struture. Clarendon Press, Oxford

48. Muthukumar M, Nickel BG (1987) J Chem Phys 86:460
49. Gupta SK, Forsmann WL (1972) Macromolecules 5:779
50. Takahashi A, Yamori S, Kagawa I, (1962) Kogyo Kagaku Zasshi 83:11
51. Nagasawa M (1988) In: Nagasawa M (ed) Studies in Polymer Science. Elsevier, Amsterdam 2:49 and references therein
52. Yamakawa H, Fujji M (1973) Macromolecules 6:407
53. Reith D, Müller B, Müller-Plathe F, Wiegand S (2002) J Chem Phys 116:9100
54. Schweins R, Huber K (2003) Polymer, submitted
55. Akcasu AZ, Benmouna M (1978) Macromolecules 11:1193
56. Barrett A (1984) J Macromolecules 17:1561
57. Huber K, Burchard W, Akcasu AZ (1985) Macromolecules 18:2743
58. Burchard W (1983) Advances in Polymer Science 48:1
59. Vrentas JS, Liu HT, Duda JC (1980) J Polym Sci Polym Phys Ed 18:633
60. Schmidt M, Burchard W (1981) Macromolecules 14:210
61. Armstrong RW, Strauss UP (1969) Encyclopedia of Polymer Science and Technology. Wiley, Vol. 10, p. 781
62. Lifson S, Katchalsky A (1954) J Polym Sci 13:43
63. Alfrey T, Berg PW, Morawetz H (1951) J Polym Sci 7:543
64. Wagner, M (2000) Ph.D.-thesis, University of Stuttgart
65. Oppermann W, Wagner M (1999) Langmuir 15:4089
66. Takahashi A, Kato N, Nagasawa M (1970) J Phys Chem 74:944
67. Karlström, G (1985) J Phys Chem 85:4962
68. Kjellander, R (2001) Electrostatic Effects in Soft Matter 317
69. Israelachvili, J (1992) Intermolecular and Surface Forces, 2nd edn. Academic Press, London
70. Kosower, Klinedinst, JACS **78**, 3493 (1956)
71. Hoffmann, RW (1976) Aufklärung von Reaktionsmechanismen. Thieme, p 129
72. Müller, RH (1996) Zetapotential und Partikelladung. WVG Stuttgart
73. Pochard I, Foissy A, Couchot P (1999) Colloid Polym Sci 277:818
74. Huber K (1993) J Phys Chem 97:9825
75. Schweins R, Huber K (2001) Eur Phys JE 5:117
76. Schweins R (2002) PhD Thesis Fachbereich 13 "Chemie und Chemietechnik" der Universität Paderborn
77. Sabbagh I, Delsanti M (2000) Eur Phys JE 1:75
78. Francois J, Truong ND, Medjahdi G, Mestdagh MM (1997) Polymer 38:6115
79. Peng S, Wu C (1999) Macromolecules 32:585
80. Heitz C, Francois J (1990) Polymer 40:3331
81. Sabbagh I, Delsanti M, Lesieur P (1999) Eur Phys JB 12:253
82. Wu C, Zhou S (1995) Macromolecules 28:5388

Received: November 2002

Polyelectrolyte Theory

C. Holm[1] · J. F. Joanny[2] · K. Kremer[1] · R. R. Netz[6] · P. Reineker[4] · C. Seidel[3] · T. A. Vilgis[1] · R. G. Winkler[5]

[1] Max-Planck-Institut für Polymerforschung, Ackermannweg 10, 55128, Mainz, Germany
E-mail: holm@mpip-mainz.mpg.de
E-mail: kremer@mpip-mainz.mpg.de
E-mail: vilgis@mpip-mainz.mpg.de

[2] Institut Curie Section Recherche, Physicochimie Curie, 26 rue d'Ulm,
75248 Paris Cedex 05, Paris, France
E-mail: jean-francois.joanny@curie.fr

[3] Max-Planck-Institut für Kolloid- und Grenzflächenforschung,
Am Mühlenberg, 14476 Golm, Germany
E-mail: seidel@mpikg-golm.mpg.de

[4] Abteilung Theoretische Physik, Universität Ulm, 89069 Ulm, Germany
E-mail: peter.reineker

[5] Institut für Festkörperforschung, Forschungszentrum Jülich, 52425 Jülich, Germany
E-mail: r.winkler@fz-juelich.de

[6] Sektion Physik, LMU Munich, Theresienstrasse 37, 80333 Munich, Germany
E-mail: netz@theorie.physik.uni-muenchen.de

Abstract In this chapter we review recent advances which have been achieved in the theoretical description and understanding of polyelectrolyte solutions. We will discuss an improved density functional approach to go beyond mean-field theory for the cell model and an integral equation approach to describe stiff and flexible polyelectrolytes in good solvents and compare some of the results to computer simulations. Then we review some recent theoretical and numerical advances in the theory of poor solvent polyelectrolytes. At the end we show how to describe annealed polyelectrolytes in the bulk and discuss their adsorption properties.

Keywords Polyelectrolytes · Poisson-Boltzmann theory · Density functional theory · Integral equations · Scaling theory · Annealed polyelectrolytes · Computer Simulations

1	Introduction..	69
2	**Stiff Polyelectrolytes**.................................	70
2.1	PB Theory, Density Functional Extensions, and Simulations....	70
2.2	Integral Equation Theory..............................	72
2.2.1	PRISM and Model....................................	72
2.2.2	Correlation Functions.................................	73
2.2.3	Effective Potential....................................	75
3	**Flexible Polyelectrolytes**.............................	76
3.1	Conformational Properties in Solution...................	77
3.1.1	Theoretical Description................................	77
3.1.2	Results..	78
3.1.3	Simulations of Debye Hückel Chains.....................	80

© Springer-Verlag Berlin Heidelberg 2004

3.2   Structure of Solution 81
3.2.1 Model 81
3.2.2 Results 81
3.2.3 Simulations of Polyelectrolyte Solutions in Good Solvent 84

4   Polyelectrolytes in Poor Solvent 86

4.1   Introduction 86
4.2   Pearl-Necklace Conformation 87
4.3   Stretching Pearl Necklaces 88
4.4   Simulations 90
4.4.1 Fluctuations 90
4.4.2 Single Chain Form Factor $S_1(q)$ 91
4.4.3 Scaling of the Correlation Length with Density 92
4.4.4 Force Extension Relation 93

5   Annealed Polyelectrolytes 94

5.1   Introduction 94
5.2   Theory of Annealed Polyelectrolytes in θ and Poor Solvents 95
5.2.1 θ and Good Solvent 95
5.2.2 Poor Solvents 97
5.3   Simulation of Weakly Charged Polyelectrolytes 98
5.3.1 Simulation Method 98
5.3.2 Rigid Rods 99
5.3.3 Flexible Chains 100
5.4   Ion Distribution around Strongly Charged Polyelectrolytes 101
5.4.1 Simulation Method 102
5.4.2 Definition of an Effective Charge 102
5.5   Weak Polyelectrolytes at Low-Dielectric Substrates 105

References 108

Abbreviations and Symbols

DH	Debye-Hückel
PB	Poisson-Boltzmann
OCP	One-Component-Plasma
PRISM	Polymer-Reference-Interaction-Site model
RLWC	Reference-Laria-Wu-Chandler closure
Z_c	counterion valence
Z_m	monomer valence
N	polymer chain length, number of repeat units
h(r)	total correlation function
g(r)	pair correlation function
$\beta =$	$(k_B T)^{-1}$
$v_{ij}^{HC}(r)$	hard core potential

l	bond length
l_p	persistence length
b	monomer size
l_B	Bjerrum length
λ_D	Debye (screening) length
κ	inverse Debye length
ρ	monomer density
ρ_c	counterion density
R_E	end-to-end distance
R_G	radius of gyration
f	fraction of charged monomers
$S(q)$	total structure factor
S_{IC}	inter-chain structure factor
S_1	spherically averaged chain form factor
ξ	correlation length
q^*	peak in the structure factor $S_{IC}(q)$
q_m	peak in the total structure factor $S(q)$
μ	chemical potential (of charges)
ξ_{el}	electrostatic blob size
v	virial coefficient
τ	reduced temperature

1
Introduction

Looking back over the last 30 years of the evolution of our general theoretical understanding of neutral polymers in solution or in melts, much has been achieved by employing modern scaling concepts [1]. In the case of polyelectrolytes, however, the very long range nature of the interaction makes the situation much more complex. The traditional separation of scales, which allows one to understand properties in terms of simple scaling arguments, does not work in many cases that well any more. Ideas and methods developed over many years for charged colloidal systems together with approaches which take the internal degrees of freedom of the chains but also the nature of the counterions and/or salt ions into account have to be combined and worked out. As is illustrated in the following for a few specific examples, significant progress has been made. However, compared to neutral polymers our understanding is still rather superficial. A direct experimental test of the theoretical concepts is quite often not possible due to idealizing assumptions in the theory but also lack of detailed control over the experimental system (e. g., chain molecular weight). Here computer simulations play a crucial role. Specific examples are given below. However, for computer simulations the same holds for the analytic theory as well as the experiment. Due to the long range nature of the interaction, the system sizes, as well as the typical accuracy of the results, are still significantly below what is standard for neutral polymers. Thus the present theory chapter can only give a snapshot picture of some recent developments/improvements made over the last few years within and around the DFG Schwerpunkt 'Polyelektrolyte'. The text also clearly demonstrates that with the newly developed means (both analytical and simulational) significant further progress can be expected over the next years.

2
Stiff Polyelectrolytes

Stiff linear polyelectrolytes can be approximated by charged cylinders. This is a relevant special case, applying to quite a number of biologically important polyelectrolytes with a large persistence length, like DNA, actin filaments or microtubules as well as to various synthetic polyelectrolytes such as those based on a poly(p-phenylene) backbone, which are the topic of a separate chapter in this book. Freezing the conformational degrees of freedom of real polyelectrolytes enables one to formulate theories which can be analytically solved, for example on a mean-field basis like the Poisson-Boltzmann (PB) approximation, which can also be formulated as a local density functional theory, within an integral equation theory approach, or simply by simulations, which we are going to review in the following subsections.

2.1
PB Theory, Density Functional Extensions, and Simulations

Within PB theory [2] and on the level of a cell model the cylindrical geometry can be treated exactly in the salt-free case [3, 4]. The Poisson-Boltzmann (PB) solution for the cell model is reviewed in the chapter in this volume on the osmotic coefficient. The PB approach can provide for instance new insights into the phenomenon of Manning condensation [5–7]. For example, the distance up to which counterions can be called condensed can be conveniently found via the inflection point in the log plot of the integrated radial distribution function P(r) of counterions [8, 9], defined as

$$P(r) = \frac{1}{\lambda} \int_b^r d\bar{r} 2\pi \bar{r} Z_c e_0 n(\bar{r}) \tag{1}$$

where b is the radius of the rod and λ its line charge density, Z_c the counterions valence, e_0 the unit charge, and n the counterion density.

The PB treatment considers the ions as a noninteracting ideal gas, and therefore neglects excluded volume and Coulomb interactions between the ions. Our general observation is that for small line charges, low densities, and small Coulomb couplings, the agreement between PB theory and the simulations of the full interacting system is rather nice. However, PB theory fails quantitatively (underestimated condensation) and qualitatively (overcharging, charge oscillations and attractive interactions) for higher densities, larger coupling strengths, and multivalent counterions/salt; see, e.g.,

Refs. [9–12]. In these case more refined theories are needed, see e.g. Ref. [13] for various approaches.

In the following we outline the method of Ref. [18] which attempts to retain the simplicity of the PB theory but also accommodates correlation effects within a **local** density approximation (LDA) where all the relevant interactions are included at the level of the free energy density. One starts out with the free energy density of the PB approach and adds an appropriate correlational correction to the mean-field free energy density. One attempt at the level of the Debye-Hückel theory (DH) is called DH plus Hole (DHH) [14]. It locally approximates the correlational contributions beyond mean-field theory by the free energy of the One Component Plasma (OCP) [15]. The classical OCP is an idealized system of N charged point-like particles of valence Z_c interacting in a neutralizing background. The OCP free energy density is a concave function which favors the development of inhomogeneities. As a result, the ions might separate into a low and a high density phase. In the case of the pure OCP, these inhomogeneities are balanced by the global electrostatic which prevents the phase separation. Unfortunately this mechanism is not present when one tries to apply the same idea to a system composed by a macroion surrounded by its ionic atmosphere. In this case, due to the non-uniform distribution of free ions, the corresponding instabilities lead to the collapse of the small ions onto the surface of the polyion what is known as "structuring catastrophe" [16, 17].

To circumvent this instability without losing the physical transparency of the method, the Debye-Hückel-Hole-Cavity DHHC theory [18] was recently proposed. In this approach, a correlation functional f_{DHHC} that is convex in density was derived. This was achieved by excluding the homogeneous background from a region of radius a around the central particle. For counterions with a diameter, this parameter a was in Ref. [18] tentatively interpreted as the ion diameter. The theory was then applied to the screening of a charged rod by its counterions. Comparisons of the ionic charge distribution obtained showed a very good agreement with the simulations for both monovalent and trivalent counterions. Figure 1 shows the Poisson-Boltzmann prediction for this observable, the results from a computer simulation [9] of the same system and the prediction of the f_{DHHC}-corrected free energy functional from Ref. [18]. The equilibrium ion distribution minimizing such a functional is most easily found using a Monte-Carlo solver [19].

It can be seen that the simulated distribution functions lie above the PB prediction, indicating an increased condensation. Moreover, this effect is more pronounced for the trivalent system. In both cases the increase is reproduced by the f_{DHHC}-corrected functional. While in the case $l_B/b=3$, $Z_c=1$ the theoretical prediction practically overlaps the simulation, it somewhat overestimates correlations in the complementary case $l_B/b=1$, $Z_c=3$. Still, the general fact that deviations from Poisson-Boltzmann theory are stronger for the multivalent system is accurately accounted for. For comparison, Fig. 1 also shows the ion distributions which would have been predicted by using the DHH correction instead. The strong deviation from the correct result is

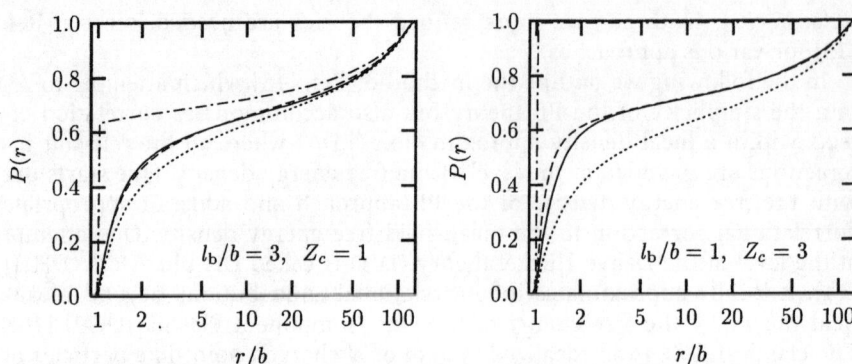

Fig. 1 Counterion distribution function $P(r)$ from Eq. (1) for two cylindrical cell models with R/b=123.8, λ=0.959 e_0/b and the values for Bjerrum length and valence as indicated in the plots. The *solid line* is the result of a molecular dynamics simulation [9] while the *dotted line* is the prediction from Poisson-Boltzmann theory. The increased counterion condensation visible in the simulation is accurately captured by the extended Poisson-Boltzmann theory (dashed line) using the DHHC correction from Ref. [18]. An approach using the DHH correction from Ref. [16] (*dash-dotted line*) evidently fails to correctly describe the ion distribution

evident. Particularly for the trivalent case the collapse of the screening atmosphere onto the macroion can clearly be seen.

2.2
Integral Equation Theory

2.2.1
PRISM and Model

The PRISM (Polymer-Reference-Interaction-Site model) theory is an extension of the Ornstein-Zernike equation to molecular systems [20–22]. It connects the total correlation function $h(r)=g(r)-1$, where $g(r)$ is the pair correlation function, with the direct correlation function $c(r)$ and intramolecular correlation functions ($\omega(r)$). For a primitive model of a polyelectrolyte solution with polymer chains and counterions only, there are three different relevant correlation functions: the monomer-monomer, the counterion-counterion, and the monomer-counterion correlation function [23, 24]. Neglecting chain end effects and considering all monomers as equivalent, we obtain the following three PRISM equations for a homogeneous and isotropic system in Fourier space:

$$h_{mm}(q) = \frac{\omega(q)}{\Delta(q)\rho_m}(1 - \rho_c c_{cc}(q) - \Delta(q)) \qquad (2)$$

$$h_{cc}(q) = \frac{1}{\Delta(q)\rho_c}(1 - \rho_m \omega(q) c_{mm}(q) - \Delta(q)) \qquad (3)$$

$$h_{mc}(q) = \frac{\omega(q)}{\Delta(q)} c_{mc}(q) \tag{4}$$

$$\Delta(q) = 1 - \rho_c c_{cc}(q) - \rho_m \omega(q) c_{mm}(q)$$
$$+ \rho_m \rho_c \omega_m(q) \left(c_{mm}(q) c_{cc}(q) - c_{mc}^2(q) \right) \tag{5}$$

where ρ_m and ρ_c denote the monomer and counterion densities, respectively, and $\omega(q)$ is the single chain structure factor. Given a closure relation, the PRISM equations can be solved. An adequate closure relation for polyelectrolyte systems is the Reference-Laria-Wu-Chandler closure (RLWC) [23, 25], which reads

$$\omega_i(r) * c_{ij}(r) * \omega_j(r) = \omega_i(r) * c_{o,ij}(r) * \omega_j(r) - \omega_i(r) * \beta v_{ij}(r) * \omega_j(r)$$
$$+ h_{ij}(r) - h_{o,ij}(r) - \ln\left(\frac{g_{ij}(r)}{g_{o,ij}(r)}\right); \quad i,j \in \{m,c\} \tag{6}$$

where the index 0 denotes reference functions obtained from a pure hard core system of the same densities with the Percus-Yevick closure. The asterisks denote convolution integrals. It should be noted that in our notation $\omega_m(q)=\omega(q)$ and $\omega_c(q)=1$. The set of coupled integral equations together with the appropriate closures is solved iteratively until convergence is achieved using a Picard iteration scheme [20].

The polyelectrolyte chains are modeled as a linear sequence of N touching hard spheres of diameter b and charge $Z_m e$. The chain model enters via the single chain intramolecular structure factor in the PRISM equations. For the considered rod-like molecules, the structure factor is known [24]. The counterions are also modeled as charged hard spheres with diameter b and charge $Z_c e$. The overall system is neutral, hence the monomer and the counterion densities obey the relation $Z_m \rho_m + Z_c \rho_c = 0$. The solvent is described as a homogeneous dielectric continuum with the dielectric constant ε. The pair interaction potential for all ionic species is given by

$$\beta v_{ij}(r) = \beta v_{ij}^{HC}(r) + Z_i Z_j \frac{l_B}{r}; \quad i,j \in \{m,c\} \tag{7}$$

where $v^{HC}_{ij}(r)$ is the hard core potential and $l_B = \beta e^2/\varepsilon$ is the Bjerrum length.

2.2.2
Correlation Functions

The following results were obtained for systems of monovalent counterions ($Z_c=-1$) and single charged monomers ($Z_m=+1$). Charge neutrality then requires $\rho_m=\rho_c$, therefore we use in the following $\rho=\rho_c$.

Polyelectrolyte solutions exhibit liquid-like order in dilute solutions which diminishes at high concentrations (cf. Figure 8 [23, 24, 26–28]. At infinite dilution $g_{mm}(r)$ has a value close to zero at small separations and

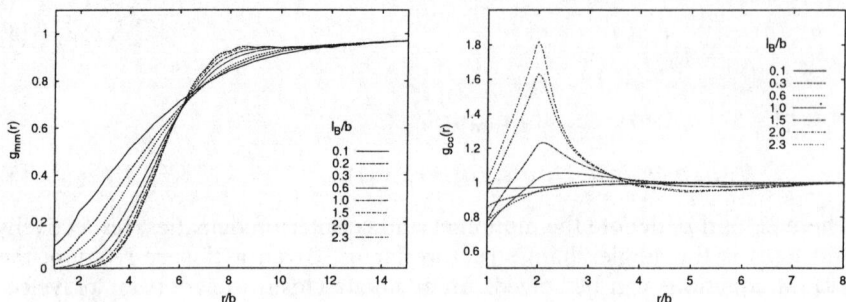

Fig. 2 Monomer-monomer ($g_{mm}(r)$, *left*) and counterion-counterion ($g_{cc}(r)$, *right*) pair correlation function for various Bjerrum lengths. The chain length is $N=80$ and the packing fraction $\eta=\pi\rho b^3/6=10^{-2}$

monotonically increases to its asymptotic value of one at larger separations. At low concentrations a peak appears on the length scale which is determined by the density of the system. As the concentration is increased further the liquid-like order first becomes more pronounced and then disappears at sufficiently high concentrations. At even higher concentrations liquid-like order appears on the length scale of the order of the size of the reference sites. For a fixed density, the ordering increases with increasing interaction strength (cf. Fig. 2 left). At very low interaction strengths the liquid-like structure is mainly determined by the hard core interaction. With increasing l_B/b the equally charged monomers repel each other and g_{mm} decreases at small distances. At the same time the structure becomes more pronounced. The characteristic peak, however, remains approximately at the same position. Thus, the characteristic length scale of the structure is mainly determined by the density. The emerging picture is similar to the behavior of simple hard sphere liquids at moderate densities.

The distribution of counterions adjacent to a monomer is captured in the monomer-counterion distribution function g_{cm} [24, 27]. For small interaction strengths ($l_B/b \ll 1$) g_{cm} is almost constant with a value of about one for all r. Hence, the local counterion density matches the bulk density almost everywhere, i.e., the counterions are distributed homogeneously over the whole system. With increasing Bjerrum length an increasing peak appears at $r=b$ and a much smaller peak at $r=2b$. Thus, the counterions are no longer homogeneously distributed, but found more likely in the vicinity of a chain. The strong increase of the counterion density next to a macroion reflects counterion condensation.

The counterion-counterion correlation function provides deeper insight into the issue of counterion condensation. As is obvious from Fig. 2 (right), g_{cc} indicates a homogeneous distribution of the counterions for small interaction strengths. With increasing l_B/b the equally charged ions repel each other and g_{cc} decreases at small distances ($r<3b$). Beyond a certain interaction strength the trend reverts and a peak appears at $r=2b$. The height of the peak increases with increasing l_B. This is the manifestation of counterion condensation. For l_B below the critical Bjerrum length for counterion condensation, an increase of l_B results in a stronger repulsion of the counterions. When the Bjerrum length crosses the critical value for Manning condensation, the counterions are attracted by the polyion and start to condense on the chain, which is accompanied by a decrease of the mean separation between ions. Hence, despite the repulsive Coulomb interaction the counterions are, for sufficiently strong interactions, subject to an effective attractive potential next to a polymer chain. The position of the peak at $r=2b$ is simple to explain if we consider the configuration with lowest energy. The electrostatic interaction forces the condensed ions to be as close to a monomer as possible but at the same time as far apart as possible from each other. This is achieved, if the counterions are located at opposite sides of the chain monomers.

2.2.3
Effective Potential

Deeper insight into the consequences of counterion condensation is gained by an effective monomer-monomer and counterion-counterion potential, respectively. The idea is to reduce the multicomponent system (macromolecules + counterions) to effective one-component systems (macromolecules or counterions, respectively). We define the simplified model in such a way that the effective potential between the counterions or monomers, respectively, of the new system yields exactly the same correlation function (g_{cc}, g_{mm}) as found in the multicomponent case at the same density. Starting from the correlation function g_{cc} –respectively g_{mm}–of the multicomponent model we calculate an effective direct correlation function c_{eff} via the one-component Ornstein-Zernike equation. An effective potential is then obtained from the RLWC closures of the one- and multicomponent models [24]. For low and moderate densities the effective potential is well approximated by

$$\beta v_{ii,\text{eff}}(r) = \beta v_{ii}(r) + \left(c_{ii}(r) - c_{ii,\text{eff}}(r)\right) \tag{8}$$

where $i \in \{m, c\}$. Hence, the effective potential is equal to the bare potential plus a modification given by the correlation functions of the multicomponent and one-component model.

Figure 3 (left) exhibits the effective counterion-counterion potential for different Bjerrum lengths. As is obvious from this figure, the effective potential is purely repulsive for low values of l_B and can be very well approximated by the bare Coulomb potential between the counterions. With increasing Bjerrum length the potential becomes negative for distances larger than a certain critical distance leading to an attractive force between two counterions. For even larger values of l_B, the effective potential exhibits a distinct minimum at a distance of about $r=2b$ in agreement with the position of the peak in g_{cc}. Figure 3 (right) shows the effective monomer-monomer potential. Similar to the counterion-counterion potential, we observe a minimum

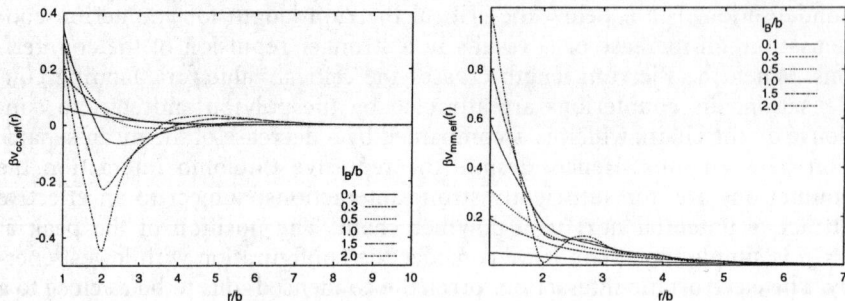

Fig. 3 Effective potential between two counterions (*left*) and two monomers (*right*), respectively for various Bjerrum lengths l_B. The chain length is $N=80$ and the density $\eta=10^{-2}$

at $r\approx 2b$ for Bjerrum lengths $l_B>1.5b$. Hence, the monomers attract each other at large interaction strengths and small distances [29–31]. The distance of $r\approx 2b$ indicates that the attraction is counterion mediated and that precisely one counterion is located between two monomers. Moreover, the effective potential is a monotonous decreasing function for $r>2b$ in agreement with the Debye-Hückel approximation. For $r<2b$ we find deviations from the Debye-Hückel potential due to the depletion interaction. The quantitative comparison between our calculations and the Debye-Hückel potential exhibits excellent agreement for the screening length as well as the dependence of the potential on the interaction strength. The deviations from the Debye-Hückel representation at short length scales of the interaction among the monomers is not surprising. The condensation of the counterions leads to a screening of the Coulomb interaction which is not captured by the Debye-Hückel potential.

Addition of salt leads to an additional screening of the Coulomb interaction [28, 32]. Our calculations for a multicomponent (polymer & counterions & salt) and a two component system (polymer & counterions) demonstrate that the correlation functions can very well be described using the Debye-Hückel potential with a salt concentration dependent screening length. The effective potentials, however, strongly depend on the salt concentration. The counterion-counterion potential is weaker, whereas the monomer-monomer potential is stronger. The latter is a consequence of salt ion condensation on the polymer chain.

3
Flexible Polyelectrolytes

Flexible polyelectrolytes exhibit conformational variations when the interaction strength (l_B) or the density of a system is changed. Hence, the structure factor is no longer an a priori known quantity but has to be determined in a self-consistent manner. This is achieved by casting the underlying multi-chain interactions into a medium-induced interaction potential among the segments of a single polymer chain. The medium-induced potential itself can be extracted conveniently using the PRISM approach. Since these approaches involve a number of approximations, which are very difficult to test directly by experiment, simulations will be of crucial importance along the way.

Before we will discuss the structure of polyelectrolyte solutions at non-zero densities, we will briefly address the conformational properties of a single chain at infinite dilution in the next section.

3.1
Conformational Properties in Solution

3.1.1
Theoretical Description

The conformational properties of an uncharged molecular chain are well described by a (discrete) semiflexible chain model [33]. The chain is comprised of mass points, each one may represent several monomers, at positions r_i ($i=0, ..., N$). The (average) length of a bond is l. The partition function of such a chain is given by

$$Z = \int \exp\left(-\lambda \sum_{i=2}^{N-1} \mathbf{R}_i^2 - \lambda_0\left(\mathbf{R}_1^2 + \mathbf{R}_N^2\right) - \frac{\mu}{2}\sum_{i=1}^{N-1}\left(\mathbf{R}_i - \mathbf{R}_{i+1}\right)^2\right) d^{3N}x \qquad (9)$$

where $\mathbf{R}_i = \mathbf{r}_i - \mathbf{r}_{i-1}$ denotes the bond vector, $\lambda = 3(1-t)/(2l^2(1+t))$, $\lambda_0 = 3/(2l^2(1+t))$, $\mu = 3t/(l^2(1-t^2))$, and $t = \langle \mathbf{R}_i\mathbf{R}_{i+1}\rangle/l^2$, i.e., t is the average cosine of the angle between two successive bond vectors. The terms with λ and λ_0 take into account the inextensibility constraint $\langle \mathbf{R}^2_i\rangle = l^2$ and represent the entropy penalty for chain stretching. The last term is the energy penalty for bending the polymer chain. In addition, we assume that the sites of the chain interact with each other via the screened Coulomb potential (Debye-Hückel potential)

$$v_{ij}^{\mathrm{DH}} = k_B T l_B \frac{e^{-\kappa|\mathbf{r}_i - \mathbf{r}_j|}}{|\mathbf{r}_i - \mathbf{r}_j|} \qquad (10)$$

κ is the inverse screening length. The corresponding partition function reads

$$Z = \int \exp\left(-\lambda \sum_{i=2}^{N-1} \mathbf{R}_i^2 - \lambda_0\left(\mathbf{R}_1^2 + \mathbf{R}_N^2\right) - \frac{\mu}{2}\sum_{i=1}^{N-1}\left(\mathbf{R}_i - \mathbf{R}_{i+1}\right)^2 \right.$$
$$\left. -\beta \sum_{i=0}^{N}\sum_{j>i}^{N} v_{ij}^{\mathrm{DH}}\right) d^{3N}x \qquad (11)$$

Excluded volume interactions can be incorporated by adding an appropriate term to the exponent. (Note, that the Lagrangian multipliers of this partition function have to be determined such that the constraints are satisfied [33].) The partition function cannot be evaluated analytically. Consequently, we have to resort to a perturbation calculation. Various perturbation schemes were suggested [34–38]. We will use the perturbation approach according to Edwards and Singh [39]. In this approach the semiflexible chain (11) is replaced by the semiflexible chain (9) with a trial stiffness parameter t_T, which is determined in such a way that the configurational properties of the original polymer agree with those of the trial chain. As a characteristic property of the polymer we consider the mean square end-to-end distance and express $\langle \mathbf{R}^2_E \rangle$ in terms of $\langle \mathbf{R}^2_E \rangle_T$ up to first order in the perturbation $(H-H_T)$, where H is the Hamiltonian of Eq. (11) and H_T is the Hamiltonian of Eq. (9) with t replaced by t_T. $\langle ... \rangle_T$ denotes the average performed with the trial Hamiltonian and $\langle ... \rangle$ denotes the average performed with the full Hamiltonian. Since we require $\langle \mathbf{R}^2_E \rangle = \langle \mathbf{R}^2_E \rangle_T$ and neglect higher order terms t_T is determined from the equation

$$\langle \mathbf{R}^2_E \rangle_T \langle H - H_T \rangle_T - \langle \mathbf{R}^2_E (H - H_T) \rangle_T = 0 \tag{12}$$

which in turn yields λ_T, λ_{0T}, and μ_T. Due to the Gaussian nature of our Ansatz for the trial Hamiltonian, all averages can be calculated analytically. The solution of Eq. (12) allows us to calculate all characteristic quantities of the reference system, because these quantities depend on the parameter t_T only. In particular, we can introduce the persistence length $l_p = l(1+t_T)/[2(1-t_T)]$.

3.1.2
Results

Flory-type free energy calculations show that the root mean square end-to-end distance of a polyelectrolyte increases linearly with the chain length at infinite dilution and without added salt [40]. Using the above perturbation theory, scaling relations at finite densities are obtained. The influence of the interaction with other polymer chains, counterions, and added salt is captured in the Debye screening length κ^{-1}.

The dependence of the root mean square end-to-end distance on molecular weight is displayed in Fig. 4 for various screening parameters κ. Two different scaling regimes are observed: For small N the mean square end-to-end distance increases proportional to N^2, corresponding to rod-like behavior [41]. The rod-like regime is most pronounced for small κ. For large N the mean square end-to-end distance increases approximately like $N^{1.2}$, typical for a short range excluded volume interaction. With increasing N the chain length exceeds the screening length and the interaction becomes short ranged (Debye-Hückel potential). Thus, we obtain the universal $N^{1.2}$ dependence of short range interactions. With increasing κ the Coulomb interaction is more and more screened and the transition to short range behavior is found at smaller chain lengths. Similar results are presented in [38] based upon the continuum version of the above semiflexible chain model.

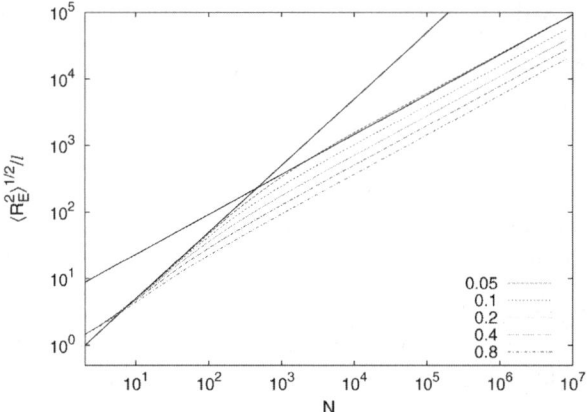

Fig. 4 Root mean square end-to-end distance of flexible polyelectrolyte chains as a function of chain length for $l_B/b=0.5$. The Debye screening length decreases from top to bottom ($\kappa=0.05, 0.1, 0.2, 0.4, 0.8$). The slopes of the straight lines are 1 and 3/5, respectively

The charge interaction leads to an increase of the persistence length of a semiflexible chain molecule [42, 43]. Using the persistence length of the reference chain, we can calculate the contribution of the electrostatic persistence length l_e to the total persistence length $l_p=l_0+l_e$. The result of our analysis is presented in Fig. 5. For $\kappa \to 0$ the electrostatic interaction is unscreened and hence l_e is independent of κ. With decreasing screening length, l_e decreases with κ. Within a certain interval l_e decays like $[\kappa b^{3/2}/l^{1/2}_B]^{-1}$. The length of the interval depends upon the interaction strength l_B/b. For $l_B/b \approx 1$, our approach yields a regime with $l_e \sim \kappa^{-2}$. To obtain the latter regime, it is important to take into account the finite chain extensibility. Similar dependencies have already been predicted by various authors [36, 37, 40, 44] and

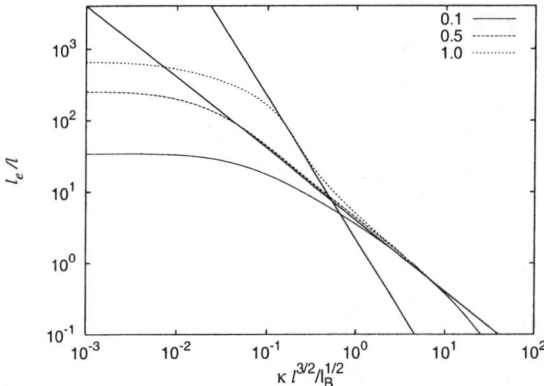

Fig. 5 Electrostatic persistence length of a flexible polyelectrolyte chain as function of $\kappa/\sqrt{l_B}$ for the Bjerrum lengths $l_B/b=0.1, 0.5, 1.0$ (bottom to top). The chain length is $N=1000$. The slopes of the straight lines are -1 and -2, respectively

are consistent with computer simulations [45, 46] for chains of finite length. Very long ($Nl\kappa \gg 1$) flexible polyelectrolyte chains exhibit Gaussian (or self avoiding walk) properties on small length scales. On an intermediate length scale the chains are linearly stretched and can be considered as a sequences of Gaussian blobs (persistent regime). Finally, at large lengths the chains are isotropically swollen, because of a short range type interaction among the blobs. The persistence length in the persistent regime is of OSF type, i.e., $l_e \sim \kappa^{-2}$, because the blobs align in a rod-like manner [37, 44]. Hence, the observed dependence $l_e \sim \kappa^{-2}$ seems to be an indication for the long chain behavior. Indeed we observe an increase of the value α, characterizing the slope of straight line $l_e \sim \kappa^{-\alpha}$, for a fixed interaction strength when we increase the chain length. A more thorough discussion will be presented in an upcoming publication.

3.1.3
Simulations of Debye Hückel Chains

A more detailed test of the currently available theoretical models is provided by various computer simulations of flexible [44, 59] and semiflexible [60] Debye Hückel chains. Here it turns out that especially the crossover towards the asymptotic regime, which is well captured by the Ansatz of Khokhlov and Khachaturian [61], gave rise to discussions over the last few years. In the case of fully flexible polyelectrolytes, weakly charged systems were in detail investigated by Micka and Kremer [59, 60]. They focused mostly on the internal structure of the chains and analyzed, in addition to the mean square end to end distance, the form factors of the chains. As it turns out, the results were inconclusive and especially for shorter chains displayed a remarkable insensitivity of the conformations as a function of the screening length. Though this did not clarify the asymptotic regime, it displayed the difficul-

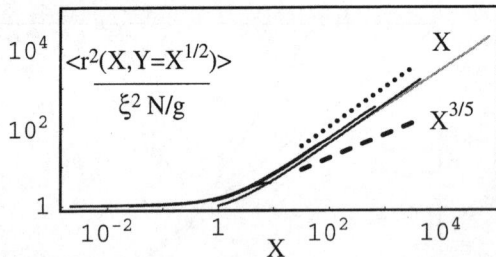

Fig. 6 Mean square extension of polyelectrolytes with Debye-Hückel interactions in a Θ-solvent relative to the corresponding uncharged chains. $X=N/g$ is the chain length measured in monomers per electrostatic blobs g. $Y=1/(\kappa\xi)$ is the screening length measured in blob diameters ξ. Data for different coupling constants are shown as a function of X, while Y is varied implicitly in such a way that the chain lengths always correspond to the electrostatic persistence length $\propto \kappa^{-2}$ predicted by Khokhlov and Khachaturian. For sufficiently strong electrostatic interactions ($X>1$) the data show the typical scaling of stiff chains, i.e. $\langle r^2 \rangle/(\xi^2 X) \propto X$. Theories predicting a shorter electrostatic persistence length scaling as κ^{-1} imply $\langle r^2 \rangle/(\xi^2 X) \propto X^{3/5}$. Adapted from [44]

ties that experiments encounter, as the effective chain lengths cover the typical experimental regime. More recently, Everaers and coworkers [44] performed a much more extensive and systematic study of the crossover from the random walk short chain regime via the "blob-rod" regime towards the self avoiding walk regime. They clearly demonstrate the validity of the picture of Khokhlov and Khachaturian and the shortcomings of the other analytical schemes. An illustration of their results is given in Fig. 6, where their data are compared to different models for the crossover towards the asymptotic regime. The downside of these findings however is, that a clear experimental demonstration will be very difficult and certainly not possible with water as solvent due to its high dielectric constant.

3.2
Structure of Solution

3.2.1
Model

To elucidate the structure of a solution of flexible polyelectrolytes, we again use the integral equation theory approach of Sect. 2.2. The necessary structure factor is determined self-consistently using the reference chain (9) of the last section. The intermolecular interactions are taken into account by a medium-induced intramolecular potential [35, 47, 48]

$$\beta W_{ij}(q) = -\rho c(q) S(q) c(q) \tag{13}$$

between two sites i and j. The effective single chain Hamiltonian is then given by (cf. Eq. (11))

$$\beta H = \lambda \sum_{i=2}^{N-1} \mathbf{R}_i^2 + \lambda_0 \left(\mathbf{R}_1^2 + \mathbf{R}_N^2 \right) + \frac{\mu}{2} \sum_{i=1}^{N-1} (\mathbf{R}_i - \mathbf{R}_{i+1})^2$$

$$+ \beta \sum_{i=0}^{N} \sum_{j>i}^{N} [v_{ij}^{DH} + W_{ij}(|\mathbf{r}_i - \mathbf{r}_j|)]. \tag{14}$$

The PRISM equations are solved iteratively using an initial guess for $W(r)$, e.g., $W(r)=0$. For the next guess the solution of the PRISM equation is used and the structure factor is determined from the solution of Eq. (12) for the reference chain. This procedure is repeated until convergence is achieved.

3.2.2
Results

The Coulomb interaction causes a chain expansion in dilute solution. With increasing density, the electrostatic interaction is screened and the chain size decreases monotonically [49, 50]. Within our model, the density of polymers are related to the inverse screening length according to $\kappa^2 = 4\pi/l_B\rho$.

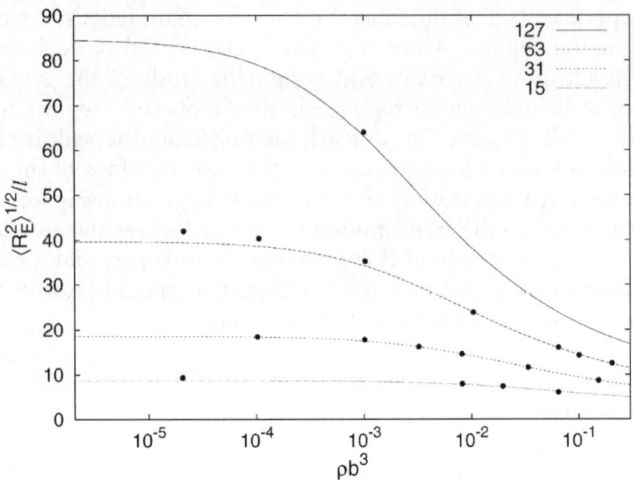

Fig. 7 Density dependence of the root mean square end-to-end distance of a flexible polymer for the chain lengths $N=127, 63, 31, 15$ (top to bottom) and $l_B/b=0.833$. Dots mark results from computer simulations of Refs. [27, 49]

Figure 7 displays the root mean square end-to-end distance as a function of the monomer density for various chain lengths. As expected, the chains are expanded in dilute solution and they contract with increasing density. The dots mark results from computer simulations [27, 49]. As is obvious from the figure, the above approach yields an excellent agreement with simulation results not only qualitatively but also quantitatively. Hence, we expect to find reliable results also for other quantities by our approach.

The monomer-monomer correlation functions of flexible polyelectrolytes exhibit qualitatively the same behavior as those for rod-like molecules. The conformational changes, however, result in more pronounced and shifted peaks. From Fig. 8 we deduce a shift of the peaks of flexible chains to larger distances compared to those of rod-like chains. This is a consequence of a smaller overlap between flexible chains compared to the one between rod-like molecules. Naturally, the effect is most pronounced for densities larger than the overlap densities. The increased peak intensity corresponds to a more pronounced order in the system of flexible chains, and is a result of the more compact structure of a polymer coil. (The structural properties of flexible polyelectrolytes without medium-induced potential have been studied in [48].)

The liquid-like order present in the pair correlation function manifests itself as a peak in the static structure factor ($S(q)$). The scaling of the position q_m of this maximum with the density has attracted much attention in the literature [40, 51–53]. Scaling arguments suggest [35, 42, 49, 51] that q_m obeys the relation $q_m \sim \rho^{1/3}$ for dilute solutions and $q_m \sim \rho^{\nu/(3\nu-1)}$ for semidilute solutions. Here ν is the scaling exponent for the end-to-end distance, i.e., $R_E \sim N^\nu$. The overlap threshold concentration is estimated as $\rho^* \sim N^{1-3\nu}$. As a conse-

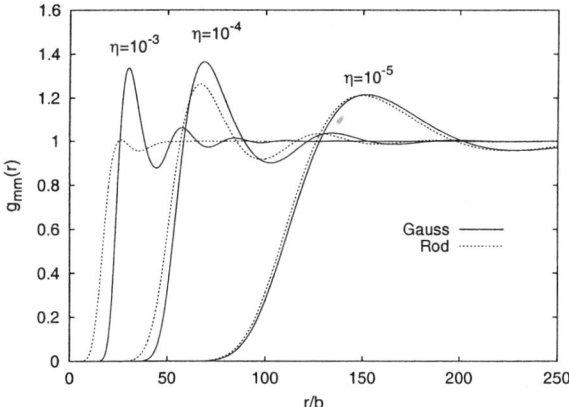

Fig. 8 Monomer-monomer ($g_{mm}(r)$) pair correlation function for the packing fractions $\eta=10^{-3}$, 10^{-4}, and 10^{-5} (left to right) of rod-like (...) and flexible chains (—). The chain length is $N=63$ and the Bjerrum length $l_B=0.5\,b$

quence, for chains in a near rod-like conformation we find $\rho^* \sim N^{-2}$. Assuming Gaussian chains, the overlap density is $\rho^* \sim N^{-1/2}$, i.e., it is much larger than for rod-like chains. In a semidilute solution the peak position of such chains should scale like $q^f_m \sim \rho$. Thus, we expect the following behavior for the peak position as a function of density: In dilute solution we observe the same dependence as for rod-like chains. Due to the conformational changes, however, we expect an extension of the dilute regime to higher values for flexible chains and chains with excluded volume interaction. Above the overlap density we should expect the scaling relations $q^f_m \sim \rho$ for Gaussian chains or $q^{ev}_m \sim \rho^{3/4}$ for chains with excluded volume interaction, respectively. In

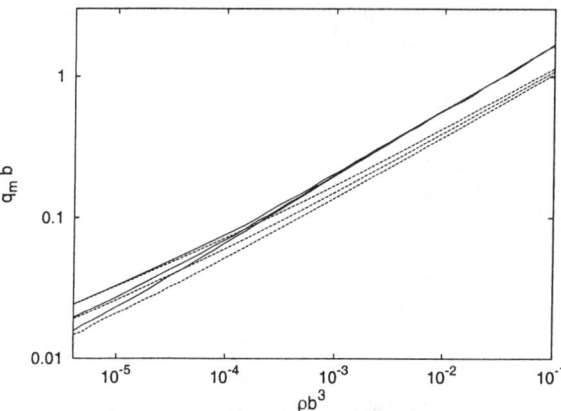

Fig. 9 Density dependence of the position of the peak, q_m, in the structure factor for flexible and rod-like (*solid lines*) chains. The chain lengths are: $N=100, 200, 500$ (from top to bottom)

between, we expect a broad transition regime since the conformations of the chains are changing. The curves of Fig. 9 confirm the expected behavior for densities below the overlap density. However, there is no transition to the scaling relation $q^f_m \sim \rho^f_m \sim \rho$ for Gaussian chains. Instead, we obtain the relation $q^f_m \sim \rho^{1/2}$ in the semidilute regime for long chains. (Our chains are to short to exhibit universal behavior. To observe the semidilute regime requires rather long chains). For shorter chains the slope is between 0.4 and 0.5. This has already been observed in [35] and is consistent with experimental results [54, 55]. Thus, the scaling argumentation does not apply. The appearance of the slope 1/2 is explained in [54] by the presence of rod-like chains of blobs. Our calculations (cf. Figure 7) and in particular the computer simulations of Ref. [49], however, suggest that the chains are not in a rod-like conformation. Hence, to understand the observed behavior requires further analytical calculations or improved scaling arguments.

3.2.3
Simulations of Polyelectrolyte Solutions in Good Solvent

So far the theoretical description focused on good solvent polyelectrolytes. Though, they are significantly simpler than poor solvent polyelectrolytes, their theoretical analysis also requires many approximations which warrant a closer look from either experiment or computer simulation. Due to the characteristic small densities in the dilute and semi-dilute regimes scattering experiments which reveal the conformational properties are very difficult to perform. Thus, simulations can help to bridge the gap between approximate analytic descriptions and experiments. Solutions of several chains with explicit counterions as well as with Debye-Hückel interactions have been performed by several authors [49, 56–58]. The first extensive study was performed by Stevens and Kremer, where chains between $N=16$ and $N=128$ charged beads with explicit counterions at various concentrations ρ and for one ρ at variable Bjerrum length l_B were investigated. Again the simulations can play a rather crucial role in checking on the general theoretical ideas. To check to what extent single chain conformations can be related to the global structure of the solution, we can compare the form factor of the single chain with the global structure function of the solution. As described below the first peak q_m in the total structure function can be related to the correlations between the different chains. Though the individual chains are never fully stretched and display, due to local fluctuations in the counterion density, significant conformational fluctuations, the scaling of the peak position q_m in the solution structure function $S(q)$ shows the expected behavior. In the dilute regime q_m varies like $\rho^{1/3}$ while at higher densities one observes the $\rho^{1/2}$ scaling, just as in experiment. The crossover density for the two power laws usually is viewed as the dilute-semidilute crossover. The mean distance of the polymer coils is approximately $2\pi/q_m$ in the dilute regime. In the semidilute regime this length denotes the blob diameter of the solution. However, it turns out that, unlike in polymer solutions of uncharged chains, this cannot that easily be related to the conformations of the chains.

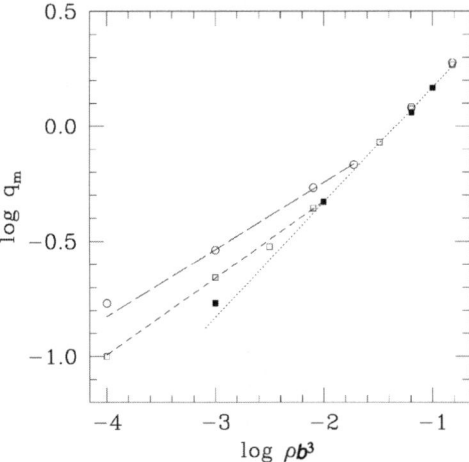

Fig. 10 Plot of the peak position q_m in the solution structure function as a function of density ρ for N=16, 32, 64. The slopes of the *dotted, dashed,* and *broken lines* correspond to 0.5, 0.33, and 0.3, respectively, from [49]

Taking the average extension of the chain $<R^2>$, the characteristic density, where spheres of that diameter would form a randomly closed packed structure, turns out to be significantly above the crossover density one finds from the q_m behavior of the scattering function, Fig. 10. The chains start to shrink long before the coils overlap, namely when the surrounding counterion clouds start to "overlap". Also the details of the conformations do no directly correspond to the simplified analysis of the solution structure factor. Taking the form factor of the chains one finds two characteristic regimes. At shorter distances the chains are significantly stretched, while at larger distances they behave like random or self avoiding walks. This lead to a form factor decay of q^{-x}, $x \leq 1$ at large q and $x=\frac{1}{\nu}$, ν=1/2, at smaller q. In the semidilute regime, without any salt present, the classical pictures identify the q_d value at the crossover with the correlation length. In the most simple pictures, which actually lead to the scaling of the collective $S(q)$, the two q values should (at least qualitatively) coincide. Figure 11 gives the characteristic q_d values as a function of the inverse screening length ($\kappa \approx \rho^{-1/2}$) for different chain lengths.

Figure 11 not only shows that there is no simple and obvious relation to the chain-chain correlation but also that the present chain lengths and densities are far from any asymptotic limit, as discussed in the section on single DH chains.

This short discussion shows that despite significant progress in the analytical theories, there are still many unsolved issues. The situation gets even more complicated if one allows for added salt and/or multivalent counterions [58] or varies the Manning ratio in the system.

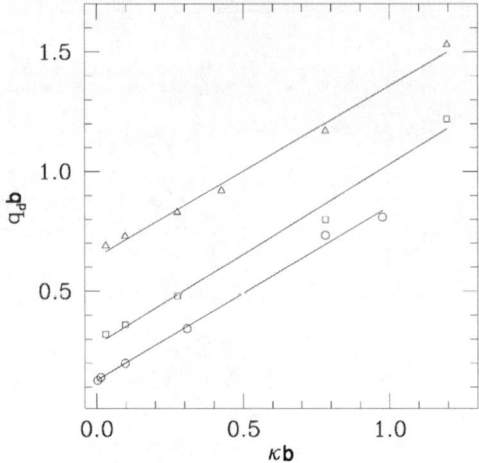

Fig. 11 Plot of the crossover q_d between extended and ideal behavior in S_q versus $\kappa \sim \rho^{-1/2}$, which shows a linear dependence. The *triangles*, *squares*, and *circles* are $N=16$, 32 and 64, respectively from [49]

4
Polyelectrolytes in Poor Solvent

4.1
Introduction

For most polyelectrolytes, the backbone is insoluble in water and the backbone monomers strongly interact through non-electrostatic attractive interactions due for example, to Van der Waals forces, hydrogen bonding or any hydrophobic effect. These hydrophobic interactions compete with the repulsive electrostatic interactions. At short distances, the hydrophobic interaction dominates and polyelectrolytes in a poor solvent are locally collapsed. At long distances, the electrostatic interactions dominate and even in a poor solvent, polyelectrolyte chains can be stretched. Although the hydrophobic interaction may have some specific character (for example hydrogen bonds are directional and there is a finite number of possible hydrogen bonds per molecule) the existing theories treat them as deriving from an isotropic short range potential characterized by a negative virial coefficient. The virial coefficient is a volume and we write it as $v=-b^3\tau$ where b is the monomer size and τ a dimensionless factor measuring the strength of the interaction.

Neutral polymers in a poor solvent (in the absence of electrostatic interactions) collapse into a dense globule [1]. The density inside the globule is $c_g \sim \tau b^{-3}$ and the size of the globule is $R_g \sim (N/c_g)^{1/3} b$. The correlation length for concentration fluctuations in the globule is $\xi_t \sim b/\tau$. The globule can for most purposes be viewed as a liquid droplet and the connectivity of the chain does not play any important role. For example the free energy of the globule is dominated by an interfacial free energy with the pure solvent with a surface tension $\gamma \sim k_B T/\xi^2_t \sim \tau^2$.

When an external force is applied to the end point of a chain in a globular conformation, the chain elongates [62]. If the force is weak, the globule is slightly deformed into an ellipsoid. When the aspect ratio of the ellipsoid is too large, (of order 2 or so) the elongated drop conformation becomes unstable, the attractive interactions are not strong enough to resist the external force and the chain discontinuously stretches to an extended conformation that would be the conformation of a Gaussian chain under the same external force. The critical force where this discontinuous transition between a collapsed globule and an extended chain occurs is $\varphi_c \sim k_B T_\tau/b$.

The first attempt to describe polyelectrolyte chains in a poor solvent was by Khokhlov [63]. He described the chain conformation as a globule strongly elongated by the electrostatic interactions which would thus look like a cigar. The diameter and the length of the cigar are imposed by a balance between surface tension and electrostatic free energies. It was however realized later that this conformation is unstable and changes to the so-called pearl-necklace conformation [64].

The next subsections describe the properties of the pearl-necklace structure and the elongation of the pearl-necklace polymer chain by an external force. We will then present numerical simulations of single polyelectrolyte chains in a poor solvent.

4.2
Pearl-Necklace Conformation

As first discussed by Rayleigh, an oil liquid droplet in water undergoes, with increasing charge, an instability when its electrostatic energy reaches the order of the interfacial energy [65]. It splits into two smaller drops. With a further increase in the charge, each daughter drop can itself undergo the same Rayleigh instability. For a drop of radius R, the critical charge at which the instability occurs is $Q_R \sim (\gamma R^3/(k_B T l_B))^{1/2}$ were l_B denotes the Bjerrum length.

A neutral collapsed polymer chain can be considered in a first approximation as a liquid drop which undergoes the Rayleigh instability when it becomes charged [64, 66]. The various daughter drops are however linked into a chain and the daughter drops cannot separate from each other. They remain linked by stretched polymer strands. The picture that is obtained for a polymer chain in a poor solvent is thus that of a necklace of collapsed globules, the pearls, connected by the strands that are stretched by the electrostatic interactions between the pearls.

The pearls are just at the Rayleigh instability threshold, their density is that of a collapsed globule and their size is obtained from the Rayleigh charge. It is the so-called electrostatic blob size

$$\xi_{el} \sim b \left(\frac{b}{f^2 l_B}\right)^{1/3} \tag{15}$$

where f is the degree of charging. The number of monomers in a pearl is $g \sim c_g \xi^3_{el} \sim \tau b f^{-2} l^{-1}_B$. The pearl necklace structure is an equilibrium between the collapsed pearls and the extended strands. The tension in the strands is therefore exactly equal to the critical force φ_c necessary to induce a transition between the two structures in a collapsed polymer chain. This tension is created by the electrostatic interactions, it is therefore of the order of the electrostatic force between neighboring pearls $\varphi \sim k_B T l_B f^2 g^2/d^2$ where d is the distance between pearls. This gives a distance between pearls $d \sim (\tau b^3 f^{-2} l^{-1}_B)^{1/2}$ and the size of the polyelectrolyte in the pearl-necklace conformation reads

$$R \sim Nf \left(\frac{bl_B}{\tau}\right)^{1/2} \tag{16}$$

It decreases with the solvent quality and crosses over to the radius of a Gaussian polyelectrolyte for $\tau \sim (f^2 l_B/b)^{1/3}$. In the pearl-necklace structure most of the polymer mass and charge belongs to the pearls but the size of the chain is dominated by the stretched strands.

It is interesting to discuss the transition between the pearl-necklace structure when there are many pearls and the collapsed polymer globule obtained either when the charge is small or when the solvent is very poor. The number of pearls $n_p = N/g \sim Nf^2 l_B/(\tau b)$ is an integer. If we for example fix the charge and decrease τ (increasing thus the solvent quality), there is only one pearl and the chain is a collapsed globule if $\tau \geq \tau_1 = Nf^2 l_B/b$. The value $\tau = \tau_1$ is the threshold value for the first Rayleigh instability where a second pearl appears. Upon further decrease of τ there is a cascade of transitions at $\tau_p = Nf^2 l_B/(pb)$ where the $(p+1)$th pearl appears. This simple discussion ignores however the fluctuations in the number of pearls which are negligible when the number of pearls is small; the transitions are then well-defined; if the number of pearls is large the thermal fluctuations in the number of pearls blurs the successive transitions which are then not well-defined.

If the pearl-necklace structure contains only a few pearls, there are always pearls at the end of the chains and these pearls are slightly larger than the inner pearls. This can be proved by doing an explicit calculation of the local electrostatic potential along the necklace very similar to that done in the following section on annealed polyelectrolytes.

4.3
Stretching Pearl Necklaces

Conformations of single polyelectrolyte chains can be probed by external forces [67, 68]. An interesting question which arises from the formation of the pearl necklace structure is the mechanical response of the chain to an external force. The simplest mean field result can be derived from the free energy proposed by [64]. However, their free energy can by strongly modified to take the electrostatic interactions between the different structural ele-

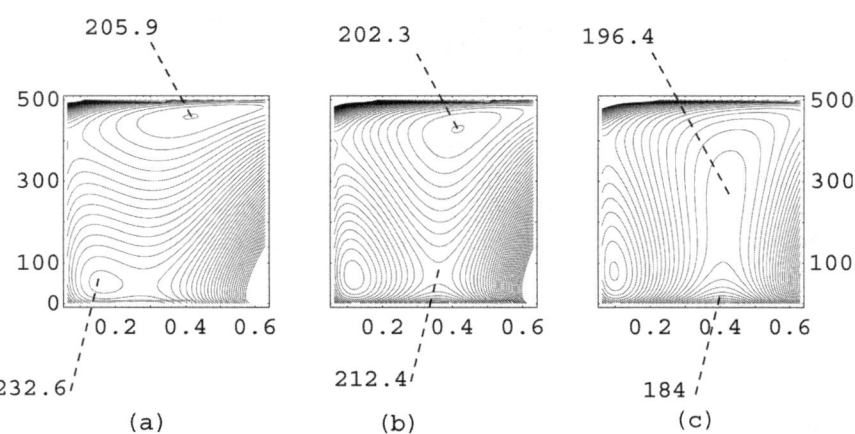

Fig. 12 Gibbs free energy in the φ,g plane for increasing applied forces. The preferred pearl size is 500. **a** In addition to the necklace minimum, an open string minimum appears (this minimum becomes deeper for higher applied forces); **b** The applied force is such that the two minima have the same depth. They are separated by a saddle point about 10 k_BT higher in energy. **c** The metastable necklace minimum disappears

ments, such as pearls and strings, on a much better basis [69]. The result of the force extension relation is then given for the case of many pearls by

$$\varphi_0(L) = k_B T \left\{ \frac{\tau}{b}\left(1 - \alpha g^{-1/3}\right) - \frac{l_B f^2 g^2}{L^2} \right\} \qquad (17)$$

Here L is the total length of the necklace chain g is the number of monomers in the pearls as before, and α a number of the order of one. It is interesting to note that the general structure of the force extension relation is mainly determined by the natural line tension in the strings which connect the pearls $\sigma \sim \tau/b$. The electrostatic repulsion between the pearls reduces the force according to the classical electrostatic term $\propto l_B f^2 g^2/L^2$. The additional term which modifies the line tension is a direct consequence of the interaction between the strings and the pearls. The pearl unwind and devolves into strings in a first order transition, where strings and pearls coexist.

The Gibbs free energy provides also a useful investigation concerning the stability of the pearl necklace structure as it was mentioned in the preceding subsection. In Fig. 12 it is shown that the relatively flat minima can be easily disturbed by moderate forces. The shift of the minima is reached over a saddle point.

The question of fluctuations in a pearl necklace consisting of discrete pearls is more challenging. A one loop renormalization will provide some insight, since the potentials are long ranged [69]. To do so, the free energy for a pearl necklace containing p pearls is expanded around its mean field solution with respect to the equilibrium number of monomers in a pearl g_0 to second order. The corresponding force part beyond the continuous one, Eq. (17) is then derived to

$$\varphi_p(L) = \varphi_0(L) - 2\frac{\tau}{b}\frac{\left(l_B f^2/\tau^3\right)^{-1/3}}{p}(N/g_0 - L/\tau b g_0 - p) \tag{18}$$

The fluctuations of the pearl number smear out the force oscillations of the pearl opening for $p>(b/l_B f^2)^{2/3}\tau$ as found previously. The fluctuations of the pearl sizes are reflected by the prefactors $p^{-1/2}(l_B f^2/b)^{-2p/3}$. As in the previous simple argument where the fluctuations have been ignored, the force shows oscillations with unstable decreasing branches that should be replaced by plateaus or pseudo-plateaus using the Maxwell construction. These results have been confirmed by detailed variational calculations [70–73].

4.4
Simulations

The predicted necklace conformations have been confirmed by simulations using only the Coulomb repulsion of the backbone charges [75], a Debye-Hückel potential [75] and by simulations using the full Coulomb potential and explicit counterions [76]. In the following we will only review the simulations using explicit counterions.

In Ref. [76] we showed that the necklace conformations can exist also in the presence of counterions and that they exhibit a variety of conformational transitions as a function of density. The end-to-end distance was found to be a non-monotonic function of concentration and showed a strong minimum in the semi-dilute regime. Here we have found for short chains a collapse of each chain into a globular stable state which repel each other due to their remaining net charge. The focus of a more recent work was to analyze, by extensive computer simulations in detail, three possible experimental observables, namely the form factor, the structure factor and the force-extension relation, which can be probed by scattering and AFM techniques [77]. The details of the simulation techniques can be found in Refs. [76, 77].

4.4.1
Fluctuations

Being first interested in single chain properties we investigated a system of several chains at a monomer density of ρ, with chain lengths between $N=100-478$. The recognition of pearls and strings was automated by a specially developed cluster algorithm [78]. The first striking observation was that the necklace conformations exhibit remarkably strong fluctuations in the size and number of pearls and strings. Analyzing the fluctuations of necklace structures we find an extended coexistence regime between different necklace structures and broad distributions for the pearl sizes and the pearl-pearl distances that tend to smear out the necklace signatures.

The coexistence regime between different structure types enlarges with increasing chain length. One reason for the small differences in the free en-

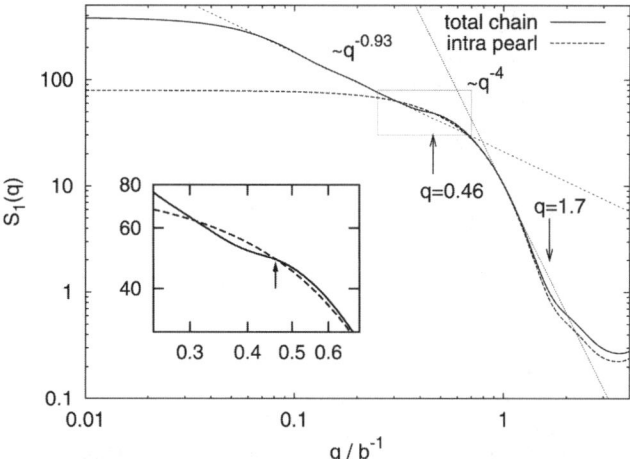

Fig. 13 Form-factor $S_1(q)$ for pearl-necklaces: (*solid*) form-factor of the whole chain, (*dashed*) form-factor due to intra pearl scattering, (*dotted*) fits. The marked region is enlarged in the *inset*

ergy is the interplay between the chain conformation and the counterion distribution. Conformations with a lower number of pearls have a smaller extension, e.g., R_E, and a larger pearl size. This leads to a stronger attraction of counterions and yields a smaller effective charge on the chain which in turn stabilizes larger pearl sizes and smaller chain extensions. In contrast to scaling theories [63, 69, 79, 80] we do not find a collapse into a globular state due to this effect. This is consistent with a more refined theoretical analysis [81] that takes prefactors and finite concentrations into account. From the probability $p(n)$ of finding a n pearl structure we calculate the free energy difference $\Delta F^{nm}=k_BT \ln (p(n)/p(m))$. For $N=430$ we find $\Delta F^{45}=-1.33k_BT$, $\Delta F^{56}=0.66k_BT$, $\Delta F^{67}=1.90k_BT$. All values are of the order k_BT which is consistent with the observed large coexistence regime. We find typically many transitions between different structure types already within one single chain during our measurement time. Not only the pearl number, also the position and size of the pearls fluctuate strongly.

4.4.2
Single Chain Form Factor $S_1(q)$

Scattering experiments can probe the conformation via the chain form factor $S_1(q)$ given by $S_1(q) = \frac{1}{4\pi N} \left\langle \sum_{i,j=1}^{N} \frac{\sin(qr_{ij})}{qr_{ij}} \right\rangle$ where $\langle ... \rangle$ denotes the ensemble average. Figure 13 shows the measured form factor for $N=382$. The radius of gyration R_G can be calculated in the Guinnier regime ($R_Gq \ll 1$) from $S_1(q)=N(1-(R_Gq^{2/3}))$, giving $R_G=16.8\ b\pm 0.3b$, in agreement with the directly calculated value $R_G=16.9b\pm 0.4b$. In the range $0.07b^{-1} \leq 0.3b^{-1}$ S_1 scales as

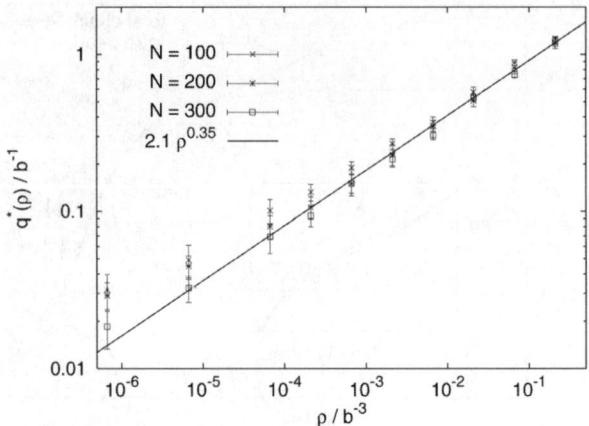

Fig. 14 Density dependence of the peak q^* in the structure factor for three different chain length N=100, 200, 300 with f=0.5. The *black line* is a fit to the data with N=200

$q^{-0.93}$ which corresponds to a stretched object. Around $q=0.46b^{-1}$ the form factor has an inflection point. A comparison with the intra pearl scattering reveals that this is due to inter pearl scattering (inset in Fig. 13). Dividing out the intra pearl form factor gives access to the inter pearl scattering and thus to the distance r_{pp} between neighboring pearls. The form factor yields r_{pp}=13.6b, in accord with the directly measured value r_{pp}=13.3b. In the region around $q=1.0b^{-1}$ we find $S_1(q) \propto q^{-4}$, the typical Porod scattering. From the small dip at $q=1.7\ b^{-1}$ one can calculate the radius r_P of the pearls to be r_P=2.6 b which again compares well to the real space value $r_P \simeq 3b$. We conclude that the cooperative effect of fluctuations on overlapping length scales broadens all characteristic signatures which can be revealed by scattering under experimental conditions (polydispersity, charge fluctuations, etc.), and necklaces might be difficult to detect in this way.

4.4.3
Scaling of the Correlation Length with Density

The overall scattering function $S(q)$ of the solution contains additional experimental information. We analyze here the inter chain scattering $S_{IC}=S/S_1$. For good solvent PEs experiments [82], theory [83], and simulations [49] we find a pronounced first peak of S_{IC} at $q^*=(2\pi)/\xi$, where ξ is the correlation length. The position varies as $q^* \propto \rho^{1/3}$ in the very dilute regime and crosses over to a $\rho^{1/2}$ regime at higher concentrations. In Fig. 14 we have plotted the density dependence of q^* in poor solvent for different chain lengths. Within the error bars we find that for poor solvent chains q^* scales proportional to $\rho^{0.35\pm0.04}$ for *all* concentrations and chain lengths. Possible reasons for this are that the response of the polyelectrolyte conformation to density changes is much larger in the poor solvent case [76, 84] than in the good solvent case

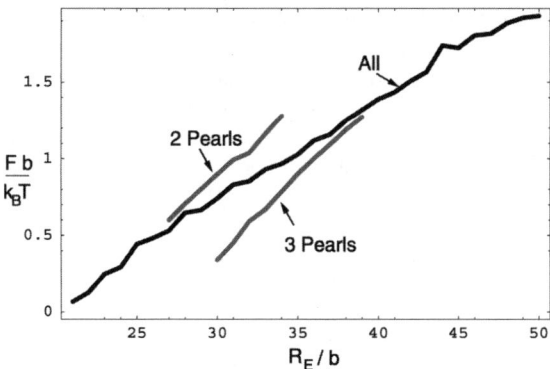

Fig. 15 Force-extension relation for a transition from two to three pearls. Shown are the average over all conformations, over only two-pearl configurations and only three-pearl configurations

[76, 84], and the chain extension behaves non-monotonic as a function of density [76, 84]. Furthermore, in the density regime between $\rho=10^{-2}b^{-3}...10^{-4}b^{-3}$ the chain extension and the pearl number varies most, and almost all monomers are located within the pearls. Upon approaching the dense regime, the string length tends to zero and we find a chain of touching pearls, indicating that the conventional necklace picture breaks down. Our results for the scaling of q^* are compatible to scaling exponents found in scattering experiments [85] and agree also very well with very recent measurements [86]. In addition we observe that the chains form a transient physical network at $\rho=0.2b^{-3}$ for $N \geq 200$ which has neither been seen in previous simulations nor predicted by theoretical approaches but is in accord with experimental studies [85]. During the simulation time these networks reconstruct several times, e.g., chains are not trapped.

4.4.4
Force Extension Relation

A different way to measure a necklace signature can be obtained by single-molecule force spectroscopy, e.g., stretching the chain by AFM [87] or optical tweezers. The case of an imposed end-to-end distance was investigated in Refs. [66, 88, 89] for a weakly charged chain at infinite dilution, for which a saw-tooth pattern in the force-extension curve was predicted. We performed simulations in the weak coupling limit, using a single chain with $N=256$, $f=1$, $l_B=0.08b$, and no counterions present. At equilibrium this system is in a two-pearl configuration with $R_E=21b$. The force-extension curve was obtained by imposing various fixed R_E up to $R_E=50b$. Then the force on the end-monomers was measured, see Fig. 15. The first remarkable observation is that the two-pearl state evolves into a three-pearl state under extension, which was predicted in [88, 89]. This is counterintuitive from simple arguments, but can be shown to result from the electrostatic inter-pearl interactions [84]. Only when the force averages where computed separately for the three- and two-pearl states, do we recognize the predicted saw-tooth pattern in the force. In equilibrium, one would expect a rounded plateau for the transition, and the upper and lower part would correspond to metastable superheated and supercooled states. Again, due to the large fluctuations all nontrivial signatures in the force-extension relation are washed out.

5
Annealed Polyelectrolytes

5.1
Introduction

In the chemistry literature, one finds the classification of "strong" and "weak" polyelectrolytes which refers to different dissociation behavior. Strong polyelectrolytes, polysalts as, e.g., Na-polystyrene sulfonate, dissociate completely in the total pH range accessible by experiment. The total charge as well as its specific distribution along the chains is solely imposed by chemistry, i.e., by polymer synthesis. In the language of statistical mechanics of disordered systems, the charge distribution is a quenched variable. That is why such polyelectrolytes are also called "quenched". On the other hand, weak polyelectrolytes, represented by polyacids and polybases, dissociate only in a rather limited pH range. The total charge on the polymer is not fixed but it can be tuned by changing the pH of the solution. The number of charges as well as their positions are fluctuating thermodynamic variables. The imposed quantity is the pH of the solution which is the chemical potential of the charges (or the field conjugate to the number of charges). In this case, the distribution of charges is an annealed variable which is the reason that in physics literature such polymers are called "annealed" polyelectrolytes.

The fraction f of charged monomers on an annealed polyelectrolyte is measured in a titration experiment. The simplest description of this experiment [90] is to assume that the polymer is homogeneous and to minimize its grand canonical free energy (its free energy at constant pH). In a mean field approach, the free energy of one chain is

$$F(f) = F_{el}(f) + k_B T N [f \log f + (1-f) \log(1-f)] - N f \mu \tag{19}$$

The first contribution is the total electrostatic free energy of the chain, the second term is the ideal gas mixing entropy for the charged and noncharged monomers along the chain and the last term is the charge chemical potential term. The charge chemical potential μ is related to the pH of the solution and the pK of the monomeric acid by $\mu = \text{pH} - \text{pK}_0$. (We have defined here these quantities with natural logarithms for simplicity). The minimization of the free energy with respect to the charge fraction f gives the titration law

$$\mu = k_B T \log \frac{f}{1-f} + \mu_{el} \qquad (20)$$

where $\mu_{el} = N^{-1} \partial F_{el}/\partial f$ is the electrostatic contribution to the chemical potential. Note that the effective pK of the polyacid is modified by the electrostatic interactions. An explicit theoretical determination of the titration curve then requires a model for the electrostatic free energy.

The distribution of the charges along an annealed polyelectrolyte chain is, however, not homogeneous due to both chain end-effects and a non homogeneous distribution of the monomers (for example in the pearl-necklace structure). This extra degree of freedom for the charges gives rise to new and non trivial features of annealed polyelectrolytes. The charge inhomogeneity has strong impact on processes dominated by end-effects, such as the self-assembly of weakly charged linear micelles [91] and adsorption on charged surfaces [92]. For end-grafted weak polyelectrolytes, a rather unusual regime has been obtained where the chain stretching (brush thickness) depends non-monotonously on salt concentration [92]. In particular, for polyelectrolytes in a poor solvent, the annealing of the charges can have strong effects [90]. The specific behavior of weak polyelectrolytes has attracted considerable interest in experimental [95–97], theoretical [90, 98–101] and simulation studies [99, 102–106].

The aim of this section is to discuss the distribution of the charges along an annealed polyelectrolyte. We first present theoretical arguments and then simulations for weakly charged polyelectrolytes. A related problem is that of strongly charged polyelectrolytes above the counterion condensation threshold [107]. The number of condensed counterions can fluctuate and they are mobile along the chain contour. A strongly charged polyelectrolyte and its condensed counterions shares thus many properties with an annealed polyelectrolyte. Simulations on the distribution of condensed counterions are presented in the last subsection.

5.2
Theory of Annealed Polyelectrolytes in θ and Poor Solvents

5.2.1
θ and Good Solvent

A weakly charged Gaussian polyelectrolyte (in a θ solvent) can be described within the electrostatic blob model as a linear array of electrostatic blobs of size ξ_{el}, given in Eq. (15). The chain being stretched, the conformational fluctuations are small, we can thus use a strong stretching approximation and describe its conformation by the average position of each monomer s along the chain, $z(s)$. The average electrostatic potential on the chain at a position z is then

$$\phi(z) = -l_B \int_{-N/2}^{N/2} ds\, f(s) \left< \frac{1}{|z-z(s)|} \right> \qquad (21)$$

where we have taken the origin at the middle of the chain. $f(s)$ is the fraction of charged monomers along the chain or the average degree of charging of monomer s. The electrostatic potential is larger at the center of the chain than close to the chain ends since it is created at the center by the two chain halves and at the end by one half chain only. The counterions feel thus a stronger attraction at the center of the chain than at the chain end and the charge fraction f of dissociated groups is larger at the chain end. For simplicity, we use here the simple electrostatic blob model with a constant chain tension. A more sophisticated description taking into account a small trumpet effect with larger blobs close to the chain end is considered in reference [101]. The electrostatic potential is in a first approximation $\phi = -l_B \langle f \rangle \xi_{el}/b^2 \log(1-(2z/R)^2)$ where $\langle f \rangle$ is the average fraction of charged monomers along the chain. The charge distribution is then obtained by writing the chemical potential balance of Eq. (20) with a varying electrostatic chemical potential $\mu_{el} = kT\phi(z)$

$$\frac{f}{\langle f \rangle} = 1 - \frac{A l_B \langle f \rangle \xi_{el}}{b^2}\left[\log(1-(2z/R)^2) + 2(1-\log 2)\right] \qquad (22)$$

where R is the chain end-to-end distance and A is an unknown pre-factor (in the definition of the electrostatic blob size). The fraction of charged monomers is indeed larger at the chain ends and decreases smoothly towards the chain center. Note that the factor $A l_B \langle f \rangle \xi_{el}/b^2$ is the Manning parameter for the condensation of the counterions on the chain of blobs and that it is small for a weakly charged polyelectrolytes. The variation of the charge along the chain is therefore a weak effect.

When the electrostatic interaction is screened, the polyelectrolyte chain is not stretched over its all contour but it is only locally stretched. The electrostatic blob model can still be used however over pieces of chains with a size of the order of the electrostatic persistence length. This length is in general larger than the Debye screening length κ^{-1}. The electrostatic potential is constant on the chain except for a small region of size κ^{-1} around the chain ends where it decreases. As in the case where the interaction is not screened the charge is higher at the chain ends but this effect is now localized over a region with a size of the order of the screening length.

For a rod-like chain, the equilibrium charge distribution was shown to obey [99]

$$\frac{f(s)}{\langle f \rangle} = 1 + \frac{\langle f \rangle l_B}{b}\left\{E_1\left(\left(\frac{N}{2}+s\right)b\kappa\right) + E_1\left(\left(\frac{N}{2}-s\right)b\kappa\right) - \frac{2}{Nb\kappa}\right\} \qquad (23)$$

where $E_1(x)$ is the exponential integral $E_1(x) = \int_x^\infty dt\, t^{-1}\exp(-t)$ [108]. The detailed calculation of the charge distribution on a flexible chain given in reference [101] leads to the same equation, but with rescaled Bjerrum and screening lengths, now expressed in terms of contour length

$$l_B \to \tilde{l}_B = l_B A \xi_{el}/b, \qquad \kappa^{-1} \to \tilde{\kappa}^{-1} = \kappa^{-1} A \xi_{el}/b \qquad (24)$$

Both distributions for rigid rods and flexible chains are compared to numerical simulations in the next section.

5.2.2
Poor Solvents

We consider a polyelectrolyte in a poor solvent in the pearl-necklace conformation. The distribution of the charges on the polymer can be calculated from the titration law (20) where the local electrostatic chemical potential is again related to the electrostatic potential created by the necklace. There are two origins for the heterogeneity of the charge distribution. One is an end-effect very similar to the one discussed for a τ solvent that we will ignore here. The other source of heterogeneity is the non-homogeneous distribution of the monomers along the chain. The monomers and thus the charges are concentrated on the pearls and the electrostatic potential is larger on the pearls. The free charged counterions therefore recombine more easily on the pearls and we expect a lower fraction of charged monomers on the pearls than on the strands.

The precise calculation of the fraction of charged monomers in the pearls is rather tedious. In order to get a first picture of the charge distribution, a two state model has been proposed in reference [101]. The monomers either belong to pearls and have a probability $\langle f \rangle - \delta f$ to be charged or they belong to the strands and they have a probability $\langle f \rangle + \delta f'$ to be charged. The average fraction of charged monomers on the chain is $\langle f \rangle$ and as explained above, we expect that both δf and $\delta f'$ are positive. As for the Gaussian polymer, the charge distribution is obtained by minimization of the grand canonical free energy of the polymer. This free energy is very similar to Eq. (19) but the polymer contribution includes both the electrostatic and the interfacial free energies of the pearl-necklace structure. In the limit where the pearls are dense $\tau \ll (f^2 l_B/b)^{1/3}$, i.e., far from the transition to the stretched Gaussian chain behavior where the pearls all disappear, the minimization of the free energy gives the reduction of the fraction of charged monomers in the pearls

$$\frac{\delta f}{\langle f \rangle} = \left(\frac{\langle f \rangle^2 l_B}{b\tau^3}\right)^{1/2} \frac{(\tau^3 l_B/b\langle f \rangle)^{1/3}}{1-(\tau^3 l_B/b\langle f \rangle)^{1/3}} \tag{25}$$

where we have ignored any numerical pre-factor. The quantity $(\tau^3 l_B/\langle f \rangle)^{1/3}$ is the electrostatic energy of a counterion at the surface of a pearl in units of $k_B T$. If this energy is smaller than $k_B T$, the charge inhomogeneity $\delta f/\langle f \rangle$ is small and the charge is almost uniform. For a finite value of this quantity, however, the denominator in Eq. (25) diverges. This is the signature of an instability of the pearl-necklace structure. As the pearl-necklace structure exists only if there are at least two pearls i.e. if $\tau \leq \tau_1 = N\langle f \rangle^2 l_B/b$, the pearl-necklace structure is stable if $\tau \leq \tau_2 \sim (b^3/N l_B^3)^{1/5}$. If the solvent quality is good enough that this condition is satisfied, the polyelectrolyte in the pearl-necklace conformation has an almost homogeneous charge distribution.

If the solvent is too poor, $\tau > \tau_2$, the pearl necklace structure becomes unstable. The polymer shows a discontinuous transition at a constant τ, upon increasing the charge, between a collapsed globular conformation and a stretched Gaussian chain conformation. This transition should be observable in a titration experiment where one plots the pH of the solution as a function of the degree of neutralization. The pH is up to a trivial additive constant the charge chemical potential $\mu(f)$. The degree of neutralization is for most practical purposes equal to the fraction of charged monomers. The titration curve is thus a plot of $\mu(f)$. These two variables being conjugate thermodynamic variables, the signature of a first order transition is a plateau in the titration curve [90].

5.3
Simulation of Weakly Charged Polyelectrolytes

Using a freely jointed bead-spring chain, the model is chosen as close as possible to that used in theory [101]. Along the chain the N monomers are connected by a harmonic potential $U_{bond} = 3k_BT/2 \sum_{n=1}^{N-1} (r_{n+1} - r_n)^2/b_0^2$, with r_n being the position of bead n and b_0 is the (bare) average bond length. N_c monomers are charged (degree of dissociation $\langle f \rangle = \langle N_c \rangle/N$) and interact via the Debye-Hückel potential $U_{DH} = kT/2 \sum_{n \neq m=1}^{N_c} l_B/r_{nm} \exp(-r_{nm}/\lambda_D)$ where the Debye screening length $\lambda_D = \kappa^{-1}$ is an input parameter. For water at room temperature, the Bjerrum length l_B which gives the strength of Coulomb interaction is about 7.1 Å. To avoid problems with counterion condensation, and to ensure that we work in a parameter range where the theory can be applied, we set the length scale by $u = l_B/b = 0.9$ with $b \approx b_0 = 1$. With this setting of the length scale, one has $b \approx 8$ Å. Hence, the polyelectrolyte chain is modeled on a coarse grained level where one bead corresponds to a few chemical monomers.

5.3.1
Simulation Method

Equilibrium properties of the polyions are investigated by Monte Carlo (MC) simulation [106]. In order to study annealed polyelectrolytes the MC simulation is performed in a semi-grand canonical ensemble where the chain is in contact with a reservoir of charges of fixed chemical potential μ. The energy change of a complete MC move is $\Delta E = \Delta E_c \pm \mu$, where ΔE_c is the change in configurational energy due to U_{bond} and U_{DH}. The plus sign is used when the monomer is to be neutralized and the minus sign when it is to be charged. Alternatively, simulations with a fixed degree of dissociation have been done in which the charges are allowed to redistribute along the chain [99].

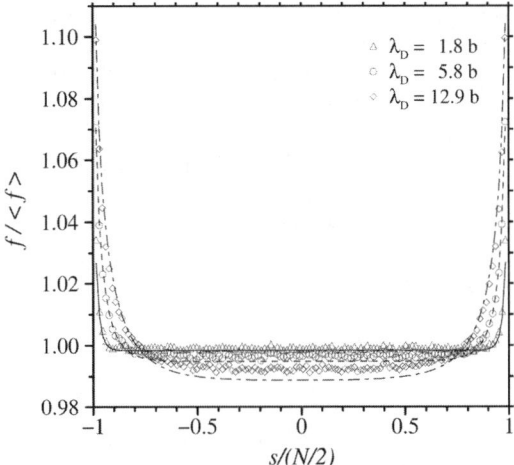

Fig. 16 Equilibrium charge distribution on a rigid rod ($N=128$, $\langle f \rangle=1/16$) at varying screening length. Theoretical results (see Eq. (23) are given as *lines*, simulation data as *symbols*

5.3.2
Rigid Rods

For rigid rods, simulation results can be compared with theoretical predictions which contain no free parameter [99]. Figure 16 shows the charge distribution at a given mean degree of dissociation for several screening

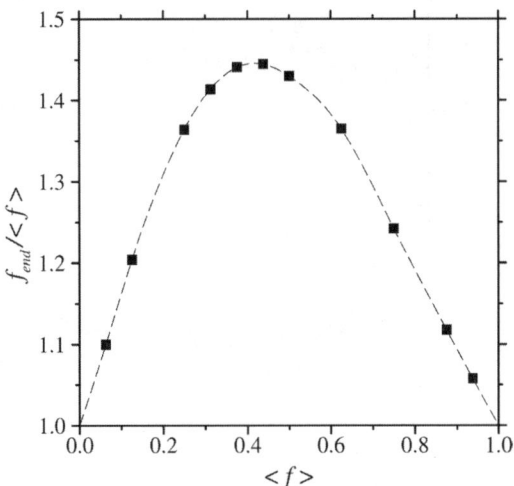

Fig. 17 Charge density at the ends versus degree of dissociation $\langle f \rangle$ ($N=128$, $\lambda_D=12.9b$). The *dashed line* is a guide for the eyes

lengths. For strong screening the agreement between theory and simulation data is quite good. For larger λ_D, finite size corrections become important. Note that the theoretical results, in particular the normalization of the charge density, is correct in the limit $N \gg \lambda_D/b \gg 1$. First order corrections in $\langle f \rangle l_B/b$ partially compensate the overestimated charge depletion in the middle of the chain seen in Fig. 16. Figure 17 gives the local charge density at the ends f_{end} for the total range of $\langle f \rangle$. The degree of charge accumulation is substantially enhanced near the maximum slightly below $\langle f \rangle = 0.5$. Note, that close to the maximum, the local charge density at the ends becomes nearly 50% higher than the average value $\langle f \rangle$.

5.3.3
Flexible Chains

Figure 18 shows the charge distributions obtained for a flexible chain together with the theoretical predictions following from Eqs. (23) and (24). We should remember that the theory contains a free parameter A, the pre-factor of the scaling theory blob size ξ_{el}. We fit A to obtain best agreement of the charge density at the middle of the chain [106]. Note that such a fitting not only corrects for the unknown pre-factor of ξ_{el}, but also for the higher order terms neglected in the theory (see the discussion above). Doing so, however, yields almost perfect agreement between simulation data and theoretical predictions for the charge accumulation at chain ends in a fairly wide parameter range ($\langle f \rangle$, λ_D). The structure of the chains can be analyzed by calculating the spherically averaged form factor

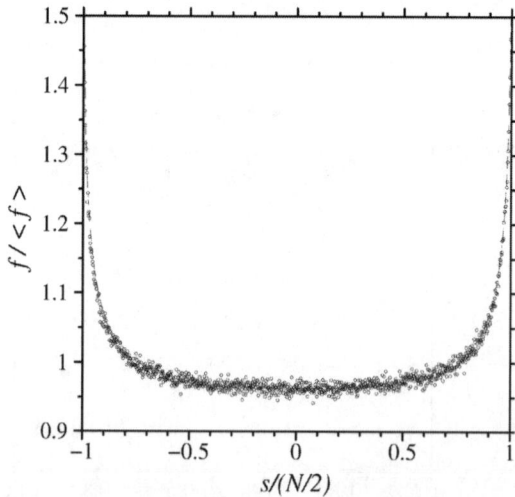

Fig. 18 Equilibrium charge distribution on a flexible chain ($N=1000$, $\langle f \rangle=0.083$, $\lambda_D=64b$). The theoretical result (see Eqs. (23, 24)) is given as *dashed line*, simulation data as *dots*

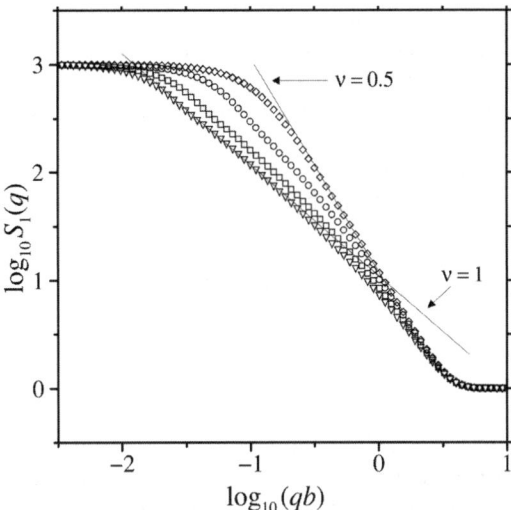

Fig. 19 Spherically averaged structure factor of annealed charged chains (N=1000): $\langle f \rangle$=0.040, λ_D=16b (*circles*); $\langle f \rangle$=0.083, λ_D 64b (*squares*); $\langle f \rangle$=0.125, λ_D=256b (*triangles*), additionally the result of an uncharged chain (*diamonds*). Thin lines indicate asymptotic scaling laws

$S_1(q) = \left\langle \left\langle 1/N \left| \sum_{n=1}^{N} \exp(iq \cdot r_n) \right|^2 \right\rangle_{|q|} \right\rangle$. From the theory of uncharged polymers we know that the structure factor scales as $S_1(q) \sim q^{-1/\nu}$ in the range $2\pi/R < q < 2\pi/b$ with ν being the universal exponent for the mean extension of the chain $R \sim N^\nu$. In Fig. 19, we plotted $S_1(q)$ for the parameter range where we obtain good agreement of theoretically predicted charge distribution and simulation data. For a not too small degree of charging and not too short screening lengths, linear scaling with $\nu \approx 1$ is reached at large length scales. On short length scales we have an almost ideal random coil behavior. This is exactly the structure implied by the theory.

5.4
Ion Distribution around Strongly Charged Polyelectrolytes

Now we will investigate end-effects in the ion distribution around strongly charged, flexible polyelectrolytes with a quenched charge distribution by molecular dynamics simulations. For sufficiently strongly charged polymers at finite densities, we always find some fraction of the counterions in close vicinity to the charged polymer backbone. These counterions effectively neutralize part of the chain charges down to some effective charge. The neutralization is stronger at those points where the electric field is stronger, since those regions attract more counterions on average. The inhomogeneous distribution of counterions lead again to an effective charge distribution along the polymer contour which has many similarities with that derived in Eq. (22) for weakly charged annealed polyelectrolytes. In a way, the positions of the counterions behave now as "annealed", and thus the distribution of all charges (fixed "quenched" monomer charges and mobile counterions) in the vicinity around the chain appears to be annealed. The delicate interplay between the electrostatic interactions, the chain conformation, and the counterion distribution has been studied in detail [107] as a function of different system parameters such as the chain length N, the charge fraction f, the charged particle density ρ_c, the ionic strength and the solvent quality, and we summarize here only some important results.

5.4.1
Simulation Method

Our model of a polyelectrolyte solution consists of N_p flexible bead-spring-chains which are located in a simulation box of length L with periodic boundary conditions. For each chain, a fraction f of the N monomers is monovalently charged ($v=1$), and fN oppositely charged monovalent counterions are added to obtain an electrically neutral system. In some cases N_s pairs of salt ions were added. The density is given in form of the charged particle density $p_c = \frac{N_p(2fN+2N_s)}{L^3}$. All ions interact via the full Coulomb potential which is computed with the P3M Ewald method [109]. We use molecular dynamics simulations in a NVT ensemble, employing a standard Langevin thermostat.

5.4.2
Definition of an Effective Charge

A snapshot of a polyelectrolyte and the counterions in its vicinity are shown in Fig. 20. Each counterion charge is associated to its closest chain monomer, and the distance to that monomer is calculated. If we add this charge to the opposite charge of the chain monomer and take the configurational average, we obtain the average local charge $q_{\text{eff}}(r,j)$ at cylindrical distance r away

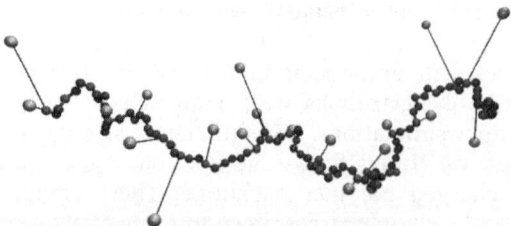

Fig. 20 The snapshot displays the geometrical situation in the vicinity of the polyelectrolyte chain. The method of measuring the distance of free ions to the chain and their assignment to individual monomers is illustrated by the connections between the ions and their closest monomer

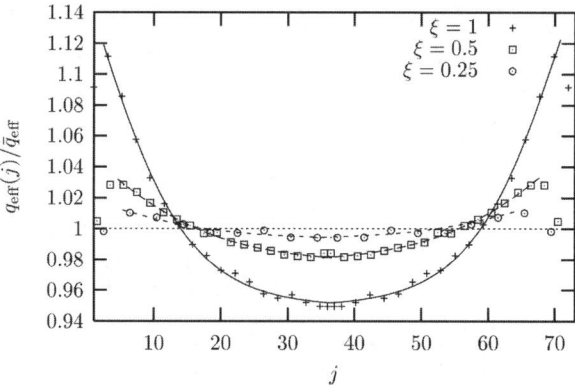

Fig. 21 Effective charge along the contour length for different values of the charge parameter ξ

from the monomer j. If we sum over all monomers we get the effective charge $q_{\text{eff}}(r)$ for the whole chain, which can be normalized to the chain charge fN via $\bar{q}_{\text{eff}} = \frac{1}{fN} q_{\text{eff}}$. We need to specify some cut-off radius r_c, up to which we want to call the ions condensed, and this is largely arbitrary. Although the quantitative results will depend on the value of r_c, the qualitative details do not change [107].

In Fig. 21 we show the dependency of $q_{\text{eff}}(j)/\bar{q}_{\text{eff}}$ for a chain of length N=72, f=1, for the values ξ=1, 0.5, 0.25. Qualitatively the curve for ξ=1 looks the same as the one in Fig. 18. For smaller ξ, the end effect vanishes, since both the number of annealing counterions and the strength of the electric field decrease strongly. In Fig. 22, we can observe the dependency of the end-effect on polymer concentration and added salt, which should be similar since in both cases the Debye length changes. In the first three cases, the

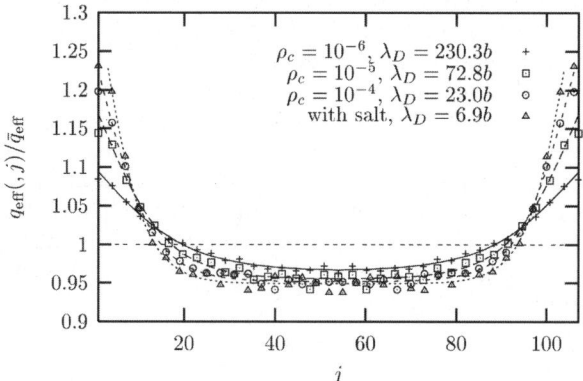

Fig. 22 Effective charge along the contour length $q_{\text{eff}}(j)$ (here with r_c=3.14b) for different densities and salt concentrations for chain length N=106, f=1/3, ξ=1

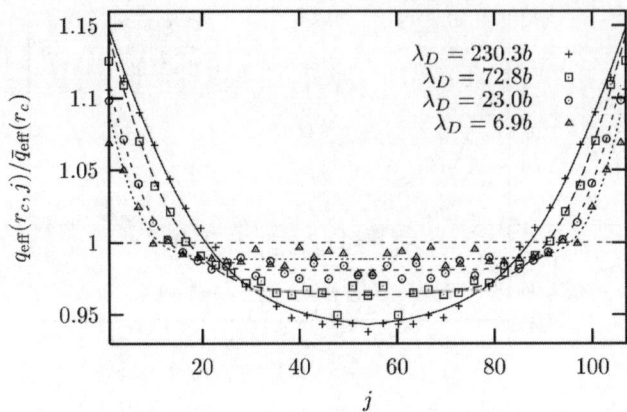

Fig. 23 Effective charge along the contour length $q_{\text{eff}}(j)$ (here with constant $\bar{q}_{\text{eff}}$=0.85) for different densities and salt concentration as in Fig. 22

increased polymer concentration (and therefore increased counterion concentration) decreases the Debye-length, and, in the last case, we add salt to decrease λ_D even further. We notice that the end effect becomes stronger with decreasing λ_D. This is opposite to the theoretical predictions and also counterintuitive, but has a simple explanation. In Fig. 23 we plotted the same data as before but this time r_c was varied for each data set to keep the average charge $\bar{q}_{\text{eff}}(r_c)$ (via the number of annealing ions) constant, because only then should the data be qualitatively compared to previously derived theoretical predictions. As can be seen in Fig. 23 we recover the expected increase of the end-effect with increasing ionic strength. If one investigates the chain length dependence of the end-effect, one finds a saturation for long chains, when the chain extension, namely R_e, is at least twice as large as the Debye screening length, suggesting that the end-effect has a penetration length of order $\approx \lambda_D$, consistent with the theoretical ideas.

Even though the chain conformation is very different in the poor solvent case, the end-effect is qualitatively the same, namely the counterions are more likely to be found at the middle of the chain than at the ends. We can also clearly see the necklace structure by looking at the effective charge along the contour length. However the string length is too short to show any charge difference in pearls and strings as has been predicted in Ref. [101]. Overall we can conclude that the charge distribution of strongly charged polyelectrolytes (with or without annealing) behaves like that one of weakly charged titrating polyelectrolytes. This is due to the presence of the mobile partially neutralizing counterions, which results in an annealed backbone charge distribution.

5.5
Weak Polyelectrolytes at Low-Dielectric Substrates

As was previously discussed, electrostatic repulsion between charges on neighboring monomers tends to decrease the effective charge of weak polyelectrolytes. This effect is stronger at low salt concentrations (i.e., for long-ranged electrostatic interactions). The situation is more complicated at dielectric boundaries, since here the charge on each monomer is interacting with its neighbors but also with its own image charge and the image charges of all its neighbors. As a result, the fraction of charged monomers decreases as the PE approaches a low-dielectric substrate (for example at a water-air interface), and even a strong PE is eventually turned into a weak PE as it comes closer to the substrate. The resulting image-charge repulsion stays finite even when the PE touches the substrate. This can explain previously puzzling experiments where poly-styrene-sulfonate (PSS, a strong PE) has been found to adsorb on the (supposedly) neutral water-air interface [110–113].

In order to treat the combined effects of added salt and dielectric boundaries on a manageable level, we use screened Debye-Hückel (DH) interactions between all charges. In the presence of a dielectric interface, the Green's function can in general not be calculated in closed form [114] except for (i) a metallic substrate (with a substrate dielectric constant $\varepsilon'=\infty$) and (ii) for $\varepsilon'=0$ (which is a fairly accurate approximation for a substrate with a low dielectric constant). For two unit charges at positions r and r' one obtains for the total electrostatic interaction including screening and dielectric boundary effects

$$v_{DH}(r,r') = l_B \frac{e^{-\kappa|r-r'|}}{|r-r'|} + l_B \frac{e^{-\kappa\sqrt{(r-r')^2+4zz'}}}{\sqrt{(r-r')^2+4zz'}} \qquad (26)$$

where z and z' denote the distance of the unit charges from the interface to the low-dielectric ($\varepsilon'=0$) half space. One sees that the second term, which is due to the dielectric interface, becomes of the same order as the first term (the ordinary bulk interaction) as one comes closer to the interface, i.e., when z and z' approach zero. Right at the interface, for $z=z'=0$, the total interaction becomes twice as large.

The ionic self energy, i.e., the interaction of one charge with its own image charge, is given by $v^{self}_{DH}(z)=v_{DH}(r,r'=r)/2$ and measures the free energetic cost of (i) immersing a single ion in an electrolyte solution, and (ii) moving this ion to a distance z from the surface. One obtains

$$v^{self}_{DH}(z) = -l_B\kappa/2 + l_B \frac{e^{-2\kappa z}}{4z} \qquad (27)$$

where the divergent Coulomb self energy has been subtracted. As a result, charges are repelled from the low-dielectric substrate.

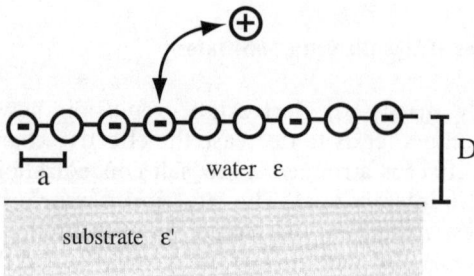

Fig. 24 A straight PE chain consisting of N dissociable monomers is placed parallel at a distance D to a dielectric interface

A PE monomer, which in all what follows is assumed to be an acid, can be either charged (dissociated) or neutral (associated), which is described by a chemical reaction $AH+H_2O \rightleftharpoons A^-+H_3O^+$ where AH denotes the associated (neutral) acidic monomer and A^- denotes the dissociated (charged) monomer. At infinite dilution, the law of mass action relates the concentrations to the equilibrium constant $K=[A^-][H_3O^+]/[AH][H_2O]$. Since the water concentration is for most purposes a constant, one defines an acid-equilibrium constant as $K_a=[A^-][H_3O^+]/[AH]$ which now has units of concentration. Defining the negative decadic logarithm of the H_3O^+ concentration and the acid constant as $pH=-\log_{10}[H_3O^+]$ and $pK_a=-\log_{10} K_a$, the law of mass action can be rewritten as $[A^-]/[AH]= 10^{pH-pK_a}$. The degree of dissociation α, defined as $\alpha=[A^-]/([AH]+[A^-])$, follows as

$$\alpha = \frac{1}{1 + 10^{pK_a - pH}} \qquad (28)$$

In the present simplified model, we neglect conformational degrees of freedom of the PE and assume a straight polymer consisting of N monomers with a bond length (i.e. distance between dissociable groups) a, located at a distance D from a dielectric interface, as is depicted in Fig. 24. This model is applicable to stiff PEs and for strongly adsorbed PEs, since they are indeed flat. The exact partition function reads

$$Z = \sum_{\{s_i\}=0,1} e^{-\mu \sum_i s_i - \sum_{i>j} s_i s_j v_{DH}(\mathbf{r},\mathbf{r}')} \qquad (29)$$

where s_i is a spin variable which is 1 (0) if the i-th monomer is charged (uncharged) and $\mathbf{r}_i$ denotes its position. The chemical potential for a charge on a monomer is given by $\mu=-2.303(pH-pK_a)+v^{self}_{DH}(D)-\ell_B\kappa/2$, the first term is the chemical free energy gained by dissociation, the second term, defined in Eq. (27), is the electrostatic interaction between the charged monomer and the substrate, and the last term is the self-energy of the released proton. All different charge distributions are explicitly summed in Eq. (29), which, together with the long-ranged interaction $v_{DH}(\mathbf{r}_i,\mathbf{r}_j)$ between charged mono-

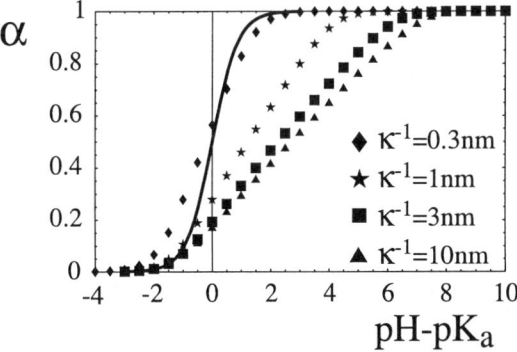

Fig. 25 Fraction of dissociated monomers α in the bulk (infinitely far away from the interface) as a function of the pH and for different salt concentrations. The polymer consists of $N=20$ monomers with a bond length $a=0.254$ nm. The *solid line* denotes the dissociation of non-interacting charges at infinite dilution

mers, defined in Eq. (26), makes the problem difficult. From the partition function, one derives the free energy per monomer, $f/k_BT=-\ln(Z)/N$, and the fraction of charged monomers as $\alpha=-\partial Z/N\partial\mu$. Previously, similar problems have been solved using mean-field theory [100, 101], restriction to nearest-neighbor repulsions only [115–117] and computer simulations [46, 106]. For the following results, the partition function is summed explicitly by an exact enumeration for a finite chain length $N=20$.

In Fig. 25 the fraction of charged monomers is presented infinitely far from the interface, i.e., in the bulk. For a screening length $\kappa^{-1}=0.3$ nm, i.e. of the order of the monomer distance $a=0.254$ nm, the electrostatic repulsion between monomers becomes irrelevant and the dissociation curve is close to the law-of-mass action Eq. (28) (note a small horizontal shift due to the increasingly negative self energy of the separated charges in the salt solution). As the salt concentration decreases, the effective charge goes down, or, as it is usually put, the apparent dissociation constant shifts. For $\kappa^{-1}=3$ nm at pH−pK$_a$=3 only half of the charges are dissociated. One notes that for screening lengths larger than the size of the polymer, $L=Na=5.1$ nm no dramatic changes are observed. As a main result, even rather strong PEs are only partially charged at low salt concentrations (where we have not taken additional complications due to chemical binding of metal ions into account [117, 118]).

In Fig. 26 results for a fixed value pH−pK$_a$=5 are shown as a function of the distance from a low dielectric substrate. Due to the strong image-charge repulsion the charge fraction α goes to zero as the distance from the substrate decreases. This is true whatever the PE strength or salt concentration, i.e., even a strong PE becomes weak at a low-dielectric substrate. The free energy exhibits a strong repulsion from the substrate, but since the charge fraction becomes zero at the substrate, the free energy stays finite all the way and is strictly bounded by zero from above. Accordingly, the maximal

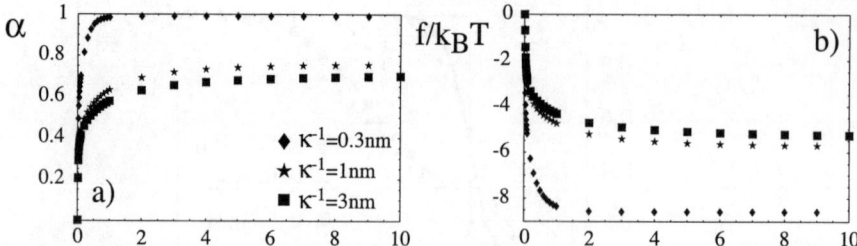

Fig. 26 Dissociation of a polymer with $N=20$ monomers as a function of the distance D from the interface to a low-dielectric half-space (idealized by $\varepsilon'=0$, see text). Shown are degree of dissociation α and free energy per monomer, f/k_BT, for fixed pH−pK$_a$=5 for different salt concentrations. As the polymer moves towards the interface the degree of dissociation and the free energy approach zero, the polyelectrolyte is repelled from the interface

free energy penalty for moving a PE from the bulk to the substrate is $f/k_BT=2.303(\text{pH}-\text{pK}_a)+\ell_B\kappa$. Since for lower salt concentration the charge fraction in the bulk goes down and the free energy is raised, while the free energy reaches zero at the interface independent of the salt concentration, one arrives at the surprising result that the repulsion from the substrate is weaker at low salt concentrations, although the electrostatic interaction by itself is more long ranged and thus more pronounced. This is a consequence of the fact that a PE is self-adjusting or regulating its degree of charge for varying salt concentrations.

These results can help to explain the recent experimental findings that strong PEs such as PSS (with roughly pK$_a$≈2) spontaneously adsorb at the water-air interface [110–113] or other low-dielectric substrates [119]. The driving force for the adsorption is the hydrophobicity of the benzene groups and the backbone, which tends to push these groups out of the water. The counteracting force is due to the charges on the sulfonic groups, which experience an image-charge repulsion from the low-dielectric half space. Naively, without taking into account the charge-regulation mechanism discussed here, one would predict the image-charge repulsion to become dominant at low salt concentration and prevent any adsorption at the interface. However, as shown here, the free energy penalty for bringing a PSS monomer (for which at neutral conditions pH−pK$_a$≈5, as used in Fig. 26) to a distance D≈0.5 nm from the interface is only a few k_BT, accompanied by a drastic decrease of the effective charge. This penalty can with ease be overcome by the hydrophobic attraction.

References

1. deGennes PG (1985) Scaling concepts in polymer physics, 2nd edn. Cornell University Press, Ithaca
2. Gouy GL (1910) Journal de Phys 9:457; Chapman DL (1913) London and Edinburg Phil Mag and J of Sci 25:475

3. Alfrey T, Berg PW, Morawetz HJ (1951) J Polym Sci 7:543
4. Katchalsky A (1971) Pure Appl Chem 26:327
5. Manning G (1969) J Chem Phys 51:924
6. Oosawa F (1971) Polyelectrolytes. Marcel Dekker, New York
7. LeBret M, Zimm B (1984) Biopolymers 23:271; ibd. 23:287
8. Belloni L, Drifford M, Turq P (1984) Chem Phys 83:147
9. Deserno M, Holm C, May S (2000) Macromolecules 33:199
10. Deserno M, Holm C, Kremer K (2001) In: Radeva T (ed) Physical Chemistry of Polyelectrolytes, vol 99, chap 2. Marcel Decker, New York, p 59
11. Deserno M, Holm C (2002) Molecular Physics 100:2941
12. Deserno M, Arnold A, Holm C (2003) Macromolecules 36:249
13. Holm C, Kékicheff P, Podgornik R (2001) Electrostatic Effects in Soft Matter and Biophysics. NATO Science Series II- Mathematics, vol 46. Physics and Chemistry, Kluwer Academic Publishers, Dordrecht
14. Nordholm S (1984) Chem Phys Lett 105:302
15. Baus M, Hansen JP (1980) Phys Rep 59:1
16. Penfold R, Nordholm S, Jönsson B, Woodward CE (1990) J Chem Phys 92:1915
17. Groot RD (1991) J Chem Phys 95:9191
18. Barbosa MC, Deserno M, Holm C (2000) Europhys Lett 52:80
19. Deserno M (2000) Physica A 278:405
20. Hansen JP, McDonald IR (1986) Theory of simple liquids. Academic Press, London
21. Curro JO, Schweizer KS (1987) Macromolecules 20:1928
22. Curro JO, Schweizer KS (1987) J Chem Phys 87:1842
23. Shew CY. Yethiraj A (1997) J Chem Phys 106:5706
24. Hofmann T, Winkler RG, Reineker P (2001) J Chem Phys 114:10181
25. Laria D, Wu D, Chandler D (1991) J Chem Phys 95:4444
26. Yethiraj A, Shew CY (1996) Phys Rev Lett 77:3937
27. Shew CY, Yethiraj A (1999) J Chem Phys 110:5437
28. Hofmann T (2002) Ph D thesis, University of Ulm, Ulm
29. Muthukumar M (1996) J Chem Phys 105:5183
30. Grønbech-Jensen N, Mashl RJ, Bruinsma RE, Gelbert WM (1997) Phys Rev Lett 78:2477
31. Ha BY, Liu AJ (1997) Phys Rev Lett 79:1289
32. Hofmann T, Winkler RG, Reineker P, in preparation
33. Winkler RG, Harnau L, Reineker P (1994) J Chem Phys 101:8119
34. Muthukumar M (1987) J Chem Phys 86:7230
35. Yethiraj A(1998) J Chem Phys 108:1184
36. Ha BY, Thirumalai D (1999) J Chem Phys 110:7533
37. Netz RR, Orland H (1999) Eur Phys JB 8:81
38. Ghosh K, Carri GA, Muthukumar M (2001)J Chem Phys 115:4367
39. Edwards SF, Singh P (1979)J Chem Soc Faraday Trans 275:1001
40. Barrat JL, Joanny JF (1996) Adv Chem Phys XCIV:1
41. Higgs PG, Orland H (1991) J Chem Phys 95:4506
42. Odijk T (1977) J Polym Sci 15:477
43. Skolnick J, Fixman M (1977) Macromolecules 10:944
44. Everaers R, Milchev A, Yamakov V (2002) Eur Phys 3 B 8:3
45. Ullner M, Jönsson S (1997)J Chem Phys 107:1279
46. Ullner M, Woodward CE (2002) Macromolecules 35:1437
47. Grayce CJ, Yethiraj A, Schweizer KS (1994) J Chem Phys 100:6857
48. Dymitrowska M, Belloni L (1999) J Chem Phys 111:6633
49. Stevens M, Kremer K (1995) J Chem Phys 103:1669
50. Winkler RG, Gold M, Reineker P (1998) Phys Rev Lett 80:373 1
51. deGennes PC, Pincus P, Velasco R, Brochard F(1976) J Phys (Paris) 37:1461
52. Förster CF, Schmidt M (1995) Adv Polym Sci 120:50
53. Hagenbuchle M, Weyerich B, Deggelmann M, Graf C, Krause R, Maier EE, Schulz SF, Klein R, Weber R (1990) Physica A 169:29

54. Nishida K, Kaji K, Kanaya T (1995) Macromolecules 28:2472
55. Essafi W, Lafuma F, Williams CE (1995) J Phys II (France) 5:1269
56. Stevens M, Kremer K (1993) Phys Rev Lett 71:2228; Macromolecules 26:4717
57. Stevens M, Kremer K (1996) J Phys 2 (France) 6:1607
58. Holm C, Kremer K (1999) Polyelectrolytes in Solution—Recent Computer Simulations. In: Noda I, Kokufuta E (eds) Proc Yamada Conf Polyelectrolytes 1998. Yamada Science Foundation, Osaka, Japan, p 27
59. Micka U, Kremer K (1996) Phys Rev B 54:2653
60. Micka U, Kremer K (1997) Europhys Lett 38:279
61. Khokhlov AR, Khachaturian KA (1982) Polymer 23:1742
62. Halperin A, Zhulina EB (1991) Europhys Lett 15:417
63. Khokhlov AR (1980) J Phys A (Math Gen) 13:979
64. Dobrynin AV, Rubinstein M, Obukhov SP (1996) Macromolecules 29:2974
65. Lord Rayleigh (1882) Philos Mag 14:184
66. Kantor Y, Kardar M (1994) Europhys Lett 27:643
67. Haronska P, Wilder J, Vilgis TA (1997) J Phys II (France) 7:1273
68. Wilder J, Vilgis TA (1998) Phys Rev B 57:6865
69. Vilgis TA, Johner A, Joanny JF (2000) Eur Phys J E 2:289
70. Migliorini G, Rostaishvili VG, Vilgis TA (2001) Eur Phys J E 4:475
71. Migliorini G, Lee N, Rostiashvili VG, Vilgis TA (2001) Eur Phys J E 6:259
72. Lee N, Vilgis TA (2002) Europhys Lett 57:817
73. Lee N, Vilgis TA (2002) Eur Phys J (in press)
74. Kantor Y, Kardar M (1995) Phys Rev B 51:1299
75. Lyulin AV, Dünweg B, Borisov OV, Darinskii AA (1999) Macromolecules 32:3264; Chodanowski P, Stoll S (1999) JChemPhys 111:6069
76. Micka U, Holm C, Kremer K (1999) Langmuir 15:4033; Micka U, Kremer K (2000) Europhys Lett 49:189
77. Limbach HJ, Holm C (2002) Computer Physics Communications 147:321; Limbach HJ, Holm C, Kremer K (2002) Europhys Lett 60:566; Holm C, Limbach HJ, Kremer K; J Phys: Condens Matter 15:205
78. Limbach HJ, Holm C (2003) J Phys Chem B, 107:8041
79. Schiessel H, Pincus P (1998) Macromolecules 31:7953
80. Dobrynia AV, Rubinstein M (1999) Macromolecules 32:915
81. Deserno M (2001) Eur Phys J E 6:163
82. Nierlich M et al. (1979) J Physique 40:701; Drifford M, Dalbiez JP (1984) J Phys Chem 88:5368; Kaji K, Urakawa H, Kanaya T, Kitamaru R (1988) 1 Physique 49:993; Wang L, Bloomfield V (1991) Macromolecules 24:5791
83. Joanny JF (2001) Scaling description of charged polymers. In: Ref. [13]
84. Limbach HJ (2001) Ph D thesis, Johannes Gutenberg Universität, Mainz, Germany
85. Essafi W, Lafuma F, Williams CE (1995) J Phys 115:1269; Heitz C, Rawiso M, François J, Polymer (1999) 40:1637; Waigh TA, Ober R, Williams CE, Galin JC (2001) Macromolecules 34:1973
86. Baigl D, Ober R, Qu D, Fery A, Williams CE (2002) Europhys Lett 63:588
87. Hugel T et al. (2001) Macromolecules 4:1039
88. Tamashiro MN, Schiessel H (2000) Macromolecules 33:5263
89. Pickett GT, Balazs AC (2001) Langmuir 17:5111
90. Raphael E, Joanny JF (1990) Europhys Lett 13:623
91. van der Schoot P (1997) Langmuir 13:4926
92. Fleer GJ, Cohen Stuart MA, Scheutjens JMHM, Cosgrove T, Vincent B (1993) Polymers at Interfaces. Chapman & Hall, London
93. Israels R, Leermakers AM, Fleer GJ (1994) Macromolecules 27:3087
94. Zhulina EB, Birshtein TM, Borisov OV (1995) Macromolecules 28:1491
95. Mandel M (1988) Polyelectrolytes. In: Mark HF, Bikales NM, Overberger CG, Menges C, Kroschwitz JI (eds) Encyclopedia of Polymers and Engineering. Wiley, New York, vol 11, p 739

96. Kötz J, Philipp B, Pfannmüller B (1990) Makromol Chem 191:1219
97. Tirtaatmadja V, Tam KC, Jenkins RD (1997) Macromolecules 30:3271
98. Jönsson B, Ullner M, Peterson C, Somelius O, Söderberg B (1996) J Phys Chem 100:407
99. Berghold G, van der Schoot P, Seidel C (1997) J Chem Phys 107:8083
100. Borukhov I, Andelman D, Borrega R, Cloitre M, Leibler L, Orland H (2000) J Phys Chem B 104:11027
101. Castelnovo M, Sens P, Joanny JF (2000) Eur Phys JE 1:115
102. Reed CE, Reed WF (1992) J Chem Phys 96:1609
103. Sassi AP, Beltrán S, Hooper HH, Blanch HW, Prausnitz JM, Siegel RA (1992) J Chem Phys 97:8767
104. Ullner M, Jönsson B, Söderberg B, Peterson C (1996) J Chem Phys 104:3048
105. Ullner M, Jönsson B (1996) Macromolecules 29:6645
106. Zito T, Seidel C (2002) Eur Phys J E 8:339
107. Limbach HJ, Holm C (2001) J Chem Phys 114:9674
108. Abramowitz M, Stegun IA (1965) Handbook of Mathematical Functions. Dover Publications, New York
109. Deserno M, Holm C (1998) J Chem Phys 109:7678; ibd 109:7694
110. Ahrens H, Förster S, Helm CA (1997) Macromolecules 30:8447
111. Klitzing R, Espert A, Asnacios A, Hellweg T, Colin A, Langevin D (1999) Coll Surf A 149:131
112. Theodoly O, Ober R, Williams CE (2001) Eur Phys J E 5:51
113. Yim H, Kent M, Matheson A, Stevens M, Ivkov R, Satija S, Majewski J, Smith GS (to be published)
114. Netz RR (1999) Phys Rev E 60:3174
115. Harris FE, Rice SA (1954) J Phys Chem 58:725 and 733
116. Marcus RA (1954) J Phys Chem 58:621
117. Lifson S (1957) J Chem Phys 26:727
118. Helm CA, Laxhuber L, Lösche M, Möhwald H (1986) Coll Pol Sci 264:46
119. Müller H, Leube W, Tauer K, Förster S, Antonietti M (1997) Macromolecules 30:2288

Received: Oktober 2002

Polyelectrolyte Complexes

Andreas F. Thünemann[1] · Martin Müller[2] · Herbert Dautzenberg[3] · Jean-François Joanny[4] · Hartmut Löwen[5]

[1] Fraunhofer Institute for Applied Polymer Research, Geiselbergstraße 69, 14476 Golm, Germany
 E-mail: andreas.thuenemann@iap.fhg.de
[2] Institute of Polymer Research Dresden, Hohe Straße 6, 01069 Dresden, Germany
 E-mail: mamuller@ipfdd.de
[3] Max Planck Institute of Colloids and Interfaces, Am Mühlenberg 1, 14476 Golm, Germany
 E-mail: dau@mpigk-golm.mpg.de
[4] Institut Charles Sadron (CNRS UPR 022), 6 rue Boussingault, 67083 Strasbourg, France
 E-mail: jean-francois.joanny@curie.fr
[5] Heinrich-Heine-Universität Düsseldorf, Universitätsstrasse 1, 40225 Düsseldorf, Germany
 E-mail: hlowen@thphy.uni-duesseldorf.de

Abstract This chapter presents selected ideas concerning complexes that are formed either by oppositely charged polyelectrolytes or by polyelectrolytes and surfactants of opposite charge. The polyelectrolyte complexes (PECs), which are surfactant-free, form typical structures of a low degree of order such as the ladder- and scrambled-egg structures. In contrast, polyelectrolyte-surfactant complexes (PE-surfs) show a large variety of highly ordered mesomorphous structures in the solid state. The latter have many similarities to liquid-crystals. However, as a result of their ionic character, mesophases of PE-surfs are thermally more stable. Both, PECs and PE-surfs can be prepared as water-soluble and water-insoluble systems, as dispersions and nanoparticles. A stoichiometry of 1:1 with respect to their charges are found frequently for both. Structures and properties of PECs and PE-surfs can be tuned to a large extent by varying composition, temperature, salt-concentration etc. Drug-carrier systems based on PECs and PE-surfs are discussed. Examples are complexes of retinoic acid (PE-surfs) and DNA (PECs). A brief overview is given concerning some theoretical approaches to PECs and PE-surfs such as the formation of polyelectrolyte multilayers.

Keywords Polyelectrolyte-surfactant complexes · Polyelectrolyte-polyelectrolyte complexes · Polyelectrolyte-colloid complexes · Polyelectrolyte-multilayers · Polyelectrolyte nanoparticles · Retinoic acid

1	Introduction	115
2	Polyelectrolyte-Polyelectrolyte Complexes (PECs)	115
2.1	Physical Background of PEC Formation	116
2.2	Water-Soluble PECs	117
2.3	Dispersions of Highly Aggregated PEC Particles	118
2.3.1	Stoichiometry of the PECs	118
2.3.1.1	Stoichiometry of Ionic Binding	118
2.3.1.1.1	Overall Composition	119
2.3.2	Structure of the PECs	119
2.3.2.1	PEC formation in pure water	119

© Springer-Verlag Berlin Heidelberg 2004

2.3.3	Effect of Salt	121
2.3.3.1	PEC Formation in the Presence of Salt.	121
2.3.3.2	Subsequent Addition of Salt	122
2.3.4	Temperature Sensitive PECs	123
2.4	Potential Applications of PECs in Solution	124
2.4.1	Polyelectrolyte-Enzyme Complexes	124
2.4.2	DNA-Polycation Complexes	125
2.4.3	PLL/Polyanion Complexes	125
2.5	Surface Modification by PECs	126
2.6	Polyelectrolyte-Multilayers	128
2.6.1	Dissociation degree of PEMs and PECs	129
2.6.2	Multilayers of PECs	130
2.6.2.1	Anisotropic Multilayers	131
2.6.2.2	Protein/PEM Interaction	133
3	**Polyelectrolyte-Surfactant Complexes (PE-surfs)**	**135**
3.1	PE-Surfs in the Solid State	135
3.2	Dispersions and Nanoparticles	136
3.3	Drug Carriers	137
3.3.1	Immobilization of Retinoic Acid by Polyamino Acids [142]	138
3.3.1.1	Chain Conformation	139
3.3.1.2	Solid-State Structures	140
3.3.1.3	Nanoparticles	141
3.3.2	Block copolymers [153]	144
3.3.2.1	Crystallinity	145
3.3.2.2	Nanostructures in the Solid State	146
3.3.2.3	Core-Shell Nanoparticles	148
3.3.2.4	Helix-Coil Transition	149
3.3.3	Polyethyleneimine [179]	152
3.3.3.1	Nanostructures	152
3.3.3.2	Thin Films	155
3.3.3.3	Release Properties	156
3.3.3.4	Nanoparticles	158
4	**Theory of Polyelectrolyte Complexation**	**161**
4.1	Debye-Hückel Theory of Polyelectrolyte Complexes	161
4.2	Polyelectrolyte Multilayers	164
4.3	Block Polyampholytes	165
4.4	Effective Interaction Between Two Polyelectrolyte Complexes	166
References		**166**

1
Introduction

An increasing number of articles on polyelectrolyte complexes reflect the growing scientific and industrial interest in this field. Some reviews are given in [1-7]. Polyelectrolyte complexes can be roughly divided into two types: The first type (PECs) are complexes of cationic and anionic polyelectrolytes. The second type (PE-surfs) are complexes of anionic polyelectrolytes and cationic surfactants and those of cationic polyelectrolytes and anionic surfactants. In its most simple form complex formation is observed when the two oppositely charged species-polyelectrolyte and polyelectrolyte or polyelectrolyte and surfactant are mixed in aqueous solution. But a number of different procedures to form PECs and PE-surfs have been developed. For example, multilayer films of PECs on solid surfaces were prepared by chemisorption from solution. This is well-known as the layer-by-layer technique and synonymously as electrostatic self-assembly [3]. The first experiments on multilayers made of oppositely charged polyelectrolytes were carried out by Decher et al. [8]. The resulting superlattice architectures of the PECs are somewhat fuzzy structures. Some reasons are i.) the build-up process of consecutive adsorption of polycations and polyanions is kinetically controlled and ii.) the polyelectrolytes are typically flexible molecules. But the absence of crystallinity in these films is expected to be beneficial for many potential applications [09]. Meanwhile the layer-by-layer method has been extended to other materials such as proteins [10, 11] and colloids (e. g. inorganic nanosheets of the clay mineral montmorillonite) [12]. Moreover, hollow nano- and microspheres are obtained via layer-by-layer adsorption of oppositely charged polyelectrolytes on template nano- and microparticles [13, 14].

The complex formation of PECs and PE-surfs is closely linked to self-assembly processes. A major difference between PECs and PE-surfs can be found in their solid-state structures. PE-surfs show typically highly ordered mesophases in the solid state [15] which is in contrast to the ladder and scrambled-egg structures of PECs [2]. Reasons for the high ordering of PE-surfs are i) cooperative binding phenomena of the surfactant molecules onto the polyelectrolyte chains [16-18] and ii) the amphiphilicity of the surfactant molecules. A further result of the cooperative zipper mechanism between a polyelectrolyte and oppositely charged surfactant molecules is a 1:1 stoichiometry. The amphiphilicity of surfactants favors a microphase separation in PE-surfs that results in periodic nanostructures with repeat units of 1 to 10 nm. By contrast, structures of PECs normally display no such periodic nanostructures.

2
Polyelectrolyte-Polyelectrolyte Complexes (PECs)

In many practical uses PEC formation takes place under conditions, where structure formation is mainly determined by the fast kinetics of this process, concealing the effects of different parameters of influence such as the mixing regime, medium conditions and macromolecular characteristics of the polyelectrolytes. The investigation of PEC formation in highly diluted aqueous solutions offers a much better chance of elucidating the general features of this process and to examine the consequences by varying the combination of polyelectrolytes and the formation conditions. After giving a brief description of the physical background of PEC formation and the basic findings regarding soluble PECs, we will focused on PEC formation in highly aggregating systems.

2.1
Physical Background of PEC Formation

The mixing of solutions of polyanions and polycations leads to the spontaneous formation of interpolymer complexes under release of the counterions. Complex formation can take place between polyacids and polybases, but also between their neutralized metal and halogenide salts. For free polyelectrolyte chains the low molecular counterions are more or less localized near the macroions, in the case of high charge densities, particularly because of counterion condensation. The driving force of complex formation is mainly the gain in entropy due to the liberation of the low molecular counterions. However, other interactions such as hydrogen bonding or hydrophobic ones may play an additional part. From the energetic point of view, PEC formation may even be an endothermic process, because of the elastic energy contributions of the polyelectrolyte chains, impeding the necessary conformational adaptations of the polymer chains during their transition to the much more compact PEC structures.

The reaction of polyelectrolyte complex formation can be described by the following equation:

$$(>-A^-c^+)_n + (>-C^+a^-)_m \Leftrightarrow (>-A^-C^+)_x + (>-A^-c^+)_{n-x}$$
$$+ (>-C^+a^-)_{m-x} + xa^- + xc^+ \quad (1)$$

where A^-, C^+ are the charged groups of the polyelectrolytes, a^-, c^+-counterions, n, m-number of the anionic and cationic groups in solution, n/m or m/n=X-molar mixing ratio, $\theta=x/n$, n<m or $\theta=x/m$, m<n, θ-degree of conversion. The degree of conversion determines whether the ionic sites of the components in efficiency are completely bound by the oppositely charged polyelectrolytes or whether low molecular counterions partly remain in the complex. Another characteristic quantity is the overall composition of the PEC structures at any mixing ratio. Even if the stoichiometry of the ionic binding is 1:1, the major component may be bound in excess, leading to an overcharging of the PEC particles.

PEC formation leads to quite different structures, depending on the characteristics of the components used and the external conditions of the reaction. As borderline cases for the resulting structures of polyelectrolyte complexes two models are discussed in the literature [19]:

- The ladder-like structure, where complex formation takes place on a molecular level via conformational adaptation (zip mechanism),
- and the scrambled egg model, where a high number of chains are incorporated into a particle.

Beside the determination of the PEC stoichiometry, the detailed characterization of the PEC structures is the main objective in understanding the effects of various parameters on the course of complex formation and the resulting properties of the PECs.

2.2
Water-Soluble PECs

Comprehensive and systematic studies on soluble PECs started with the pioneering work of the groups of Kabanov [20-22] and Tsuchida [23-25]. They could show that under appropriate salt conditions, PEC formation between polyions with weak ionic groups and significantly different molecular weights in non-stoichiometric systems results in soluble complexes. Such PECs are structured according to the ladder model, consisting of hydrophilic single-stranded and hydrophobic double-stranded segments. Predominantly carboxylic groups containing polyanions were used in combination with various polycations. The presence of a small amount of salt enables rearrangement and exchange processes and shifts the reaction more to thermodynamic equilibrium, leading to a uniform distribution of the short chain components among all long chains of the counterpart. Stop flow measurements [26] showed that PEC formation takes place in less than 5 µs, nearly corresponding to the diffusion collision of the polyion coils. The further addition of salt leads at first to a shrinking of the PECs due to the shielding of their charges by the electrolyte. When a critical salt concentration is exceeded, a disproportionation of the short guest chains occurs, leading to completely complexed, precipitating species and pure host polyelectrolyte chains in solution [27-29]. At still higher salt concentration the precipitate dissolves again and both components exist as free polyelectrolyte chains in solution. Similar effects can be induced by changes of the p-H-value [21].

While at low ionic strength the PECs possess a high stability, they are able to take part in polyion exchange and substitution reactions at higher ionic strengths. Especially, the addition of a component of higher molecular mass or stronger ionic groups results in substitution reactions [30, 31]. Therefore, one should always be aware that polyelectrolyte complexes are "living systems", which may respond very sensitively to changes of their environment.

2.3
Dispersions of Highly Aggregated PEC Particles

The preparation of soluble PECs requires special conditions. Most of the practical applications do not meet such demands. Particularly, PEC formation between strong polyelectrolytes results in highly aggregated or macroscopic flocculated systems. However, in extremely diluted solutions the aggregation process stops on a colloidal level, offering the possibility to study these dispersions of PEC particles in detail. We focused our interest on the investigation of the stoichiometry and structure of such PECs, looking especially for the general effects of various macromolecular and external parameters.

2.3.1
Stoichiometry of the PECs

2.3.1.1
Stoichiometry of Ionic Binding

The first step in determining the composition of the PEC particles is to find the stoichiometry of the ionic binding, i.e. of the degree of conversion. Well-established techniques of studying the endpoint stoichiometry of PEC formation are turbidity, potentiometry, conductivity, electrophoretic light scattering, and colloidal titration. A maximum or a breakpoint of the measured quantity indicate the endpoint of the reaction of complex formation (for details of these methods see Review [32]). For strong polyelectrolytes in most cases a 1:1 endpoint stoichiometry was found. Some of these techniques also allowed us to determine the stoichiometry as a function of the molar mixing ratio X of the oppositely charged components. Micheals et al. [33] studied PEC formation between sodium poly(styrene sulfonate) (NaPSS) and poly(4-vinylbenzyl-trimethylammonium chloride) (PVBTACl) by conductometry. A comparison of the conductivity of the PEC solutions with that of control solutions containing the same amount of low molecular counterions according to a 1:1 reaction and the remaining excess polyelectrolyte, proved that the component reacted completely under full release of the low molecular counterions. The same conclusion could be drawn from UV spectroscopy and potentiometry with a chloride sensitive electrode for the UV-active polyanion NaPSS in combination with poly(diallyldimethylammonium chloride) (PDADMAC) or its copolymers with acrylamide [34]. These investigations were carried out using deionized water as a solvent. NMR studies on the same polycations in combination with different NaPSS samples revealed, for NaPSS of higher molecular masses, a full binding, especially at higher ionic strength [35]. The full binding of the component in deficiency even for strong mismatching of the charge densities of the components, suggests a more smoothed charge neutralization than a strictly located salt binding.

2.3.1.1.1
Overall Composition

While for the ionic binding the 1:1 stoichiometry seems to be the rule in the case of strong polyelectrolytes ζ-potential measurements by electrophoretic light scattering [36, 37] revealed a strong overcharging for non-stoichiometric mixing ratios. In [38], viscometry was used to determine the total composition of the PEC particles between NaPSS and DADMAC-acrylamide copolymers. The stoichiometric factor decreases from ~1.5 at low mixing ratios to 1 at 1:1 mixing of charged groups. Similar results were obtained by analyzing the complex solutions and the supernatants by analytical ultracentrifugation using an UV-detector [39]. Also chromatographic techniques [40, 41] and gel electrophoresis [42] were employed in this regard.

2.3.2
Structure of the PECs

2.3.2.1
PEC formation in pure water

At first PEC formation in pure water will be considered. Electron and X-ray absorption microscopy showed that PEC formation leads to polydisperse systems of nearly spherical particles [32, 34]. To study the PEC structure in detail static and dynamic light scattering were used. Static light scattering in its traditional manner of data analysis provides information about the weight average of the particle mass and the z-average of the square of the radius of gyration. A more detailed interpretation of the shape of the scattering functions by comparison with model calculations for various structures [43-45] yields additional information about the structure type and polydispersity and allows the calculation of the structural density (reciprocal of the degree of swelling). Dynamic light scattering yields, by the cumulant fit of the correlation function the z-average of the diffusion coefficient and via the Einstein-Stokes equation, the hydrodynamic radius (for details see [46]). The ratio between the radius of gyration and the hydrodynamic radius is a structure sensitive parameter and varies from ρ_s=0.775 for monodisperse spheres to values much higher than unity for elongated structures. A Laplace transform of the correlation function directly provides the distribution function of the decay constants, and for spheres the distribution of radii [47].

The findings of light scattering studies on PEC formation between a variety of different polyanions and polycations [32, 34, 48-50] can be summarized as follows:

PEC formation in a concentration range of the component solutions below $1 \cdot 10^{-3}$ g/mL resulted in stable dispersions of PEC particles when non-stoichiometric mixing ratios are used. In general, the scattering functions of the PECs could be well fitted by the model of polydisperse systems of homogeneous spheres, using a logarithmic distribution to describe the polydispersity of radii. Taking into account the polydispersities obtained by static light scattering, the results of dynamic light scattering confirmed nicely the model used. Together with the findings about PEC stoichiometry this leads to the conclusion that the sphere-like PEC particles consist of a charge neutralized core, surrounded by an electrostatically stabilizing shell of the excess component.

The structural parameters of the PECs (mass, size, structure density) changed only slightly with the molar mixing ratio X up to about X=0.9. Mass and size of the complexes decreased somewhat in such a way that the structure density remained nearly constant. The decrease in PEC particle mass can be explained by the consumption of the excess component. Therefore, not a growing of the PEC particles, but the generation of new particles is the dominating process with increasing mixing ratio. Secondary aggregation and macroscopic flocculation occurred when the 1:1 mixing ratio is approached.

Even in extremely diluted systems ($<1\times10^{-5}$ g/ml) the PEC particles consist of several hundred single polyelectrolyte chains. The level of aggregation increases strongly with the rising concentration of the component solutions up to several thousand chains per particle [50].

For strong polyelectrolytes and suitable charge densities of the components the structure density of the PECs is high and ranges normally from 0.3 to 0.7 g/mL, indicating very compact structures. Stronger mismatching of the charge distances of the components results in higher degrees of swelling [34, 48, 49].

By contrast to expectations (from the thermodynamic point of view [51]), PEC formation between PDADMAC and a series of NaPSS of different molecular masses (8000 to 1million g/mol) did not reveal systematic changes of the structural parameters with variation of the molecular masses of NaPSS. Most likely, in pure water PEC formation is mainly governed by the kinetics of this process, leading to frozen structures, which are far from thermodynamic equilibrium. This statement is supported by the findings on preferential binding from a mixture of polyanions during the addition of a PDADMAC solution [52, 53]. An equal binding of both kinds of polyanions or even a preferential binding of the thermodynamically handicapped component (weaker ionic groups or lower molecular weight) was found. Especially, a strong favoring of a low molecular mass NaPSS in competition with a high molecular mass NaPSS was observed in pure water. Only an appropriate amount of NaCl in the solutions shifted the preferential binding to the thermodynamically favored component.

2.3.3
Effect of Salt

2.3.3.1
PEC Formation in the Presence of Salt

Because it weakens the electrostatic interaction and enables rearrangement processes salt should play a decisive part in the formation of highly aggregated PECs similar to the role of salt in the formation of soluble PECs. This topic has been addressed in detail in [54, 55, 56, 57]. The presence of a small amount of NaCl led to a dramatic decrease of the level of aggregation for PECs between NaPSS and PDADMAC by nearly two orders of magnitude. Higher ionic strengths caused secondary aggregation and again a strong increase in PEC masses and sizes. Therefore, the level of aggregation can also be controlled by the amount of salt during complex formation. In contrast to PEC formation in pure water, a remarkable increase in the aggregation level with a rising mixing ratio X was observed. PEC formation between NaPSS and a DADMAC-acrylamide copolymer with 47 mol% DADMAC in the presence of NaCl resulted in a strongly swollen PEC structures up to a mixing ratio X=0.5, where a collapse to highly aggregated and compact particles occurred. Very recent studies on PEC formation between NaPSS and well-defined copolymers between DADMAC and *N*-methyl-*N*-vinyl-acetamide (MVA) of various compositions revealed the same pattern, as shown in Fig. 1. While at 75 mol% of DADMAC complex formation is very similar to the case of PDADMAC, for the sample with 25 mol% DADMAC a strong

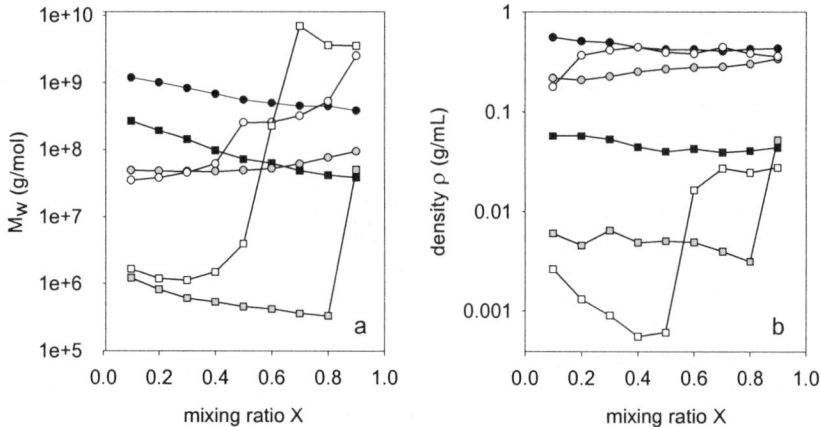

Fig. 1 Particle mass M_w and structure density ρ of PECs between DADMAC-MVA-copolymers (M_w=100 kDa) and NaPSS (M_w=66 kDa) in dependence on mixing ratio X:1 (●) copolymers with 75 mol% DADMAC, 2 - (■) copolymer with 25 mol% DADMAC, *black* - pure water, *grey* - 0.01 N NaCl, *white* - 0.1 N NaCl

increase in particle mass and structure density was found in 0.1 N NaCl at X=0.5.

2.3.3.2
Subsequent Addition of Salt

Comprehensive studies of various PEC systems [48, 55-058] revealed that the response of polyelectrolyte complexes to the addition of NaCl is be very different, strongly depending on the used polyelectrolyte components. The general tendencies may be summarized as follows:

PECs between strong polyelectrolytes shows secondary aggregation up to macroscopic flocculation (systems 1 and 2 in Fig. 2). The colloidal salt stability decreases drastically with an increasing mixing ratio. This can easily be understood by the corresponding decrease in the thickness of the stabilizing shell of the excess component and in the degree of overcharging. For the complex NaPSS/PDADMAC dissolution did not occur up to an ionic strength of 4 mol/L. The structure density remained nearly constant. By contrast, PECs of polyanions with carboxylic groups swelled and dissolved completely at a critical salt concentration (PEC 3 in Fig. 2), where this salt concentration depends on the charge densities of the components.

The colloidal salt stability can be improved by using double hydrophilic block or graft copolymers with neutral water-soluble chains [59-62], as shown in PEC 4 in Fig. 2. While, for a stabilization of the PEC particles during complex formation in pure water close to the 1:1 mixing ratio, short PEG chains (2 kDa) are sufficient, long blocks were found to be necessary to stabilize PECs against salt induced aggregation. In the first case the lowering

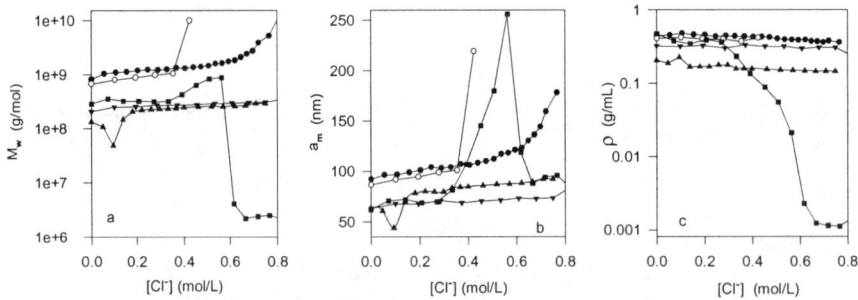

Fig. 2 Response of various PECs (first polyelectrolytes in initial solutions) to subsequent addition of salt: **a** particle mass M_w, **b** particle radius a_m (corrected for polydispersity, obeying the relation $M_w=(4\pi/3)\, \rho\, a_m^3$), **c** structure density ρ: 1(●) - NaPSS/PDADMAC, X=0.3, 2(○) - NaPSS/PDADMAC, X=0.6, 3(■) - DADMAC-acrylamide copolymer (47 mol% DADMAC)/NaPMA X=0.6, 4(▼) - DHP2 (block copolymer of poly(2-acrylamido-2-methyl-1-propanesulfonic acid) and poly(ethylene glycol), PEG block length 10 kda)/PDADMAC, X=0.6, 1-4: addition of NaCl, 5(▲) - DADMAC-acrylamide copolymer (47 mol% DADMAC)/Na-PMA, X=0.6, addition of $CaCl_2$

of the flocculation rate of the PEC particles enables the passage of this critical region during the short time of the experiment.

For recording the correct scattering curves during salt induced aggregation, it was necessary to wait up to one hour after the dosage step before the scattering intensities became sufficiently stable. Especially at low ionic strengths the process of secondary aggregation is slow.

Using a DAWN EOS multiangle light scattering instrument additionally equipped with a cylindrical cell with plan-parallel windows as a matching bath, time-resolved measurements were carried out [63]. The time dependent changes were detected after the addition of different amounts of NaCl to a complex between Na-poly(methacrylate) (NaPMA) and PDADMAC. In 0.1 N NaCl the aggregation process was completed only after 20 hours, while in 0.2 N NaCl the final state was reached in half an hour when the PEC began to precipitate. However, the dissolution process above the critical NaCl was completed in fractions of a second.

Very slow long-term changes were also observed in the investigations of the exchange processes of low molecular mass for high molecular mass NaPSS or NaPMA for NaPSS [53]. A complete exchange took place after 2 months in the presence of 0.1 N NaCl.

The occurrence of such long-term changes in salt-containing systems has to be taken into account in all applications of polyelectrolyte complexes.

Very interesting results were obtained in studies on the effect of multivalent added salts ($CaCl_2$, $MgCl_2$, $AlCl_3$, and FeCl3) [63]. NaPMA was used as the polyanion to estimate both the colloidal stability and the stability of the ionic binding. If the multivalent ion is the counterpart of the component in excess strong secondary aggregation took place, obviously caused by additional complexation of the carboxylic groups of the stabilizing shell. By contrast, PECs with NaPMA in deficiency were stabilized by multivalent cations (PEC 5 in Fig. 2). Secondary aggregation is suppressed to a great extent, but also the stability of the ionic binding is strongly enhanced. Most likely, also in this case complex formation between carboxylic groups and the cations takes place, leading to a stabilization of the core of the PEC particles and a strengthening of the stabilizing shell by additional free PDADMAC chains anchored in the core. This hypothesis were able to be confirmed by first NMR studies, which proved the binding of a part of the cations and the liberation of a part of the polycationic groups. Detailed studies are in progress.

2.3.4
Temperature Sensitive PECs

Poly(*N*-isopropylacrylamide) (PNIPAM) is the most studied thermosensitive polymer in aqueous media. It is soluble in water at low temperatures but becomes insoluble when the temperature is increased above a certain temperature (~32 °C) (lower critical solution temperature), which is related to the coil-to-globule transition [64, 65]. In the case of a polymer network, a volume change occurs reversibly within a narrow temperature range. The properties of such microgels can be varied to a great extent by the introduction of ionic groups [66-69]. Thermosensitive gels have mostly been prepared by covalent cross-linking. However, polyelectrolyte complex formation between ionically modified temperature sensitive polymers should offer a promising new route in the preparation of tailor-made gel particles on a nanometer scale. Recent studies [70] on PEC formation between ionically modified PNI-PAM has demonstrated that nearly monodisperse, sphere-like complex particles in a 100 nm scale can be prepared, which show a temperature controlled, reversible swelling-deswelling behavior by nearly a factor 10 at about 35 °C.

2.4
Potential Applications of PECs in Solution

Polyelectrolyte complexes can be prepared in a desired range of mass, size and structure density. The behavior of the PECs can be controlled by external parameters such as the ionic strength, the pH of the medium or the temperature. Therefore, such complexes should be of great interest as potential carrier systems for drugs, enzymes, or DNA because charged species can easily be integrated into the complex particles.

2.4.1
Polyelectrolyte-Enzyme Complexes

Comprehensive investigations have been carried out about the complex formation between proteins and polyelectrolytes (see reviews [71-73]). Kabanov described in [22, 74] a protection-activation mechanism for enzymes by a pH induced solution-precipitation process of their complexes with polyelectrolytes. The direct coupling of an enzyme to a polyelectrolyte often leads to a strong loss of its activity. Therefore, in [74] it was proposed to immobilize enzymes in a complex matrix of synthetic polyelectrolytes. The basic idea consists of the fact that PEC formation can be carried out under optimal pH conditions for incorporation of an enzyme into the complex and then the pH can be adjusted to the pH value for the highest activity without its release from the PEC matrix. The immobilization procedures were optimized regarding the pH, nature and molecular mass of the components as well as the mass ratio enzyme/polyelectrolytes for trypsin [75], lipases [76] and amyloglucosidase [77], yielding activities of the immobilized enzymes between 50 and 90% of the uncomplexed enzyme. In [77] the complex solution was additionally entrapped in poly(vinylalcohol) gel particles, enabling an easy handling of the enzyme in industrial applications and repeated uses without loss of activity.

2.4.2
DNA-Polycation Complexes

More than 6000 diseases are known to be caused by a single gene defect. While classical medicine can only cure the symptoms, gene therapy offers the possibility of correcting such defects by introducing DNA, or their fragments, into cells. One of the challenging problems in this respect is the development of appropriate carrier systems which guarantee a high rate of cell transfection, but a low mortality of the cells. A good survey of this area of research is given in [78]. DNA or oligonucleotide complexes with polycations have proven to be promising gene delivery systems [79, 80]. A serious problem results from their low colloidal stability, particularly near a 1:1 mixing ratio of anionic to cationic groups and under physiological salt conditions. The use of double hydrophilic polycations improved the situation decisively [81-86].

However, in many cases the high level of aggregation of the complexes remained a great disadvantage, especially for oligonucleotides, because it causes the introduction of several thousands DNA fragments into one cell. Carrying out PEC formation under conditions according to the generation of soluble PECs, the aggregation level could be drastically reduced [87].

2.4.3
PLL/Polyanion Complexes

Polypeptide/polyelectrolyte complexes are interesting for fundamental research on biomolecular recognition on the conformation level and for applications such as pharmaceutical drug carriers. Several non-biogenic polyanion solutions were mixed in excess with Poly(L-lysine) (PLL) solutions at a constant pH=6 to give negatively charged nonstoichometric dispersed PEC, i.e., all PLL is expected to be consumed by the polyanion. At this pH prior to mixing PLL adopts the random coil conformation. By CD spectroscopy it was possible to probe the influence of several polyanions on its conformation. In Fig. 3 CD spectra of selected PECs composed of PLL and the polyanions poly(vinyl sulfate) (PVS), poly(styrenesulfonate) (PSS), poly(acrylic acid) (PAC), poly(maleic acid-*co*-ethylene) (PMA-E) and poly(maleic acid-*co*-α-methylstyrene) (PMA-MS) are given. Generally, a negative CD band around 190 nm can be diagnostically assigned to random coil conformation and the 222/206 nm doublet (n–π^*, π–π^* transitions) to the α-helical conformation of polypeptides. The latter could be seen in the PLL complexes with PAC, PMA-E and PVS. In contrast PMA-MS and PSS showed similar CD spectra like that of the uncomplexed PLL at pH=6. No definite rule could be derived from these findings, since the strong and supposedly more extended PVS induces the α-helix in contrast to the also strong PSS. Furthermore the weak and less extended PAC and PMA-E (the head/head adduct of PAC with different charge distances) tended to be α-helix formers. Similar conformation directing results were found for PLL/PAC complexes by Shinoda [88], who claimed a stoichiometric reaction between PLL and PAC and

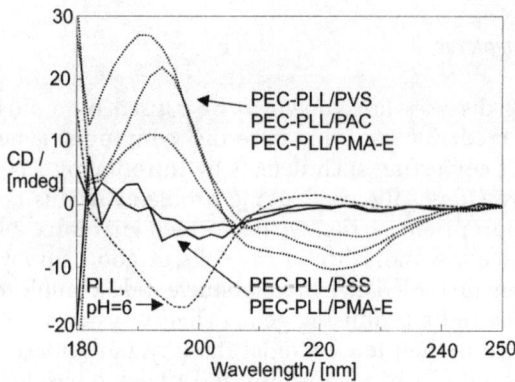

Fig. 3 CD spectra of negatively charged nonstoichometric PECs consisting of PLL and of different polyanions (PA, indicated in the plot) mixed at n−/n+=2.0 (c_{PLL}= c_{PA}=0.001m, pH=6) and of PLL at pH=6

furthermore a left-handed super helix of PAC around the right-handed α-helix of PLL. Since both PMA-MS and PSS did not influence the PLL conformation at pH=6, it might be concluded, that aromatic ring containing polyanions cannot bind PLL in the α-helical state due to preassociated structures of these polyanions.

2.5
Surface Modification by PECs

Surface modification by polyelectrolytes and PECs has been established for many years [89, 90]. For example, the interaction between the oppositely charged polymers poly(diallyldimethylammonium chloride) and poly(maleic acid-*co*-methylstyrene) in the presence of cellulose leads to a strong surface modification [91]. Model systems that are based on PEC nanoparticles with a small size distribution were used for the investigation of their adsorption on different surfaces [92]. The deposition of PEC particles on silica surfaces has been characterized by in-situ-ATR-FTIR, which has been shown to be suitable for monitoring adsorption processes at the liquid/solid interface on the molecular level [93], and SEM methods [94]. In Fig. 4 there are representative in-situ-ATR-FTIR spectra on the adsorption of PECs consisting of PDADMAC/PMA-MS (n−/n+=0.6) on a silicon substrate, which has been premodified by a six layered polyelectrolyte multilayer assembly of poly(ethyleneimine) (PEI) and poly(acrylic acid) (PAC). Prominent IR bands of the PEC like the $\nu(CH)$ and the $\nu(COO^-)$ evolve with increasing adsorption time paralleled by the incline of the negative $\nu(OH)$ band, which is due to the adsorption induced subsequent water removal from the surface. The integrals of these bands are given in Fig. 5 showing that the adsorption kinetics of the PECs is rather slow and takes more than two hours for completion of the surface coverage. A representative SEM image of these adsorbed PEC-

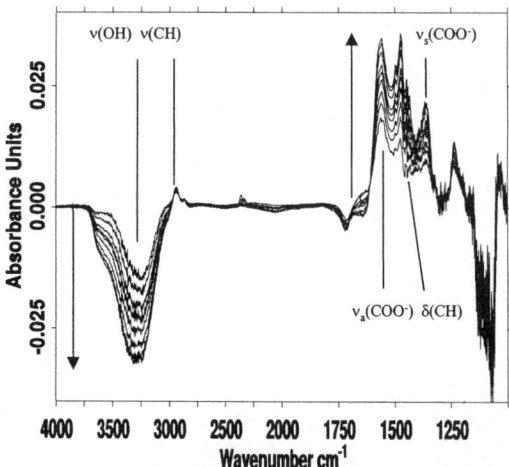

Fig. 4 in-situ-ATR-FTIR spectra monitoring the adsorption of PEC particles composed of PDADMAC/PMA-MS-0.6 at a negatively charged multilayer modified surface in dependence on adsorption time (c_{PEC}=0.0032m, pH=6)

particles composed of PDADMAC and PMA-MS on the negatively charged surface is given in Fig. 6. There, individually adsorbed droplet-like particles in the range of 200-1000 nm with a certain distribution maximum around 450 nm were observed with slight tendencies to fusion on the surface [95]. The shape of these particles appeared to be hemispherical, whereas that of

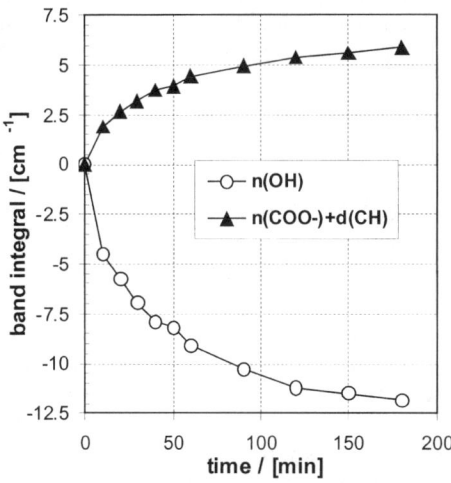

Fig. 5 Integrated areas of the ν(OH) band (water) and the composed band of ν(COO$^-$) and ν(CH) (PEC particles) (from Fig. 4) as a linear scale for PEC surface coverage in dependence of adsorption time

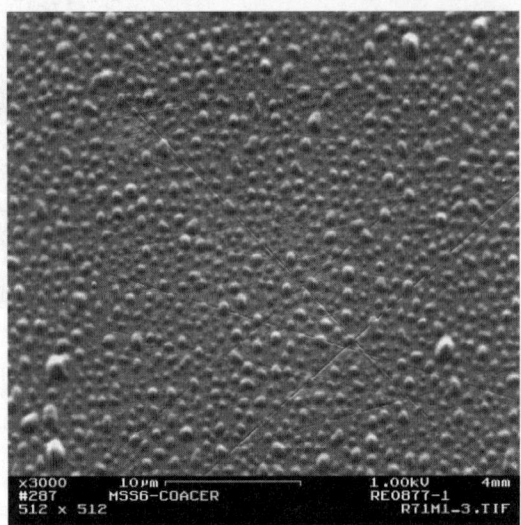

Fig. 6 Adsorbed positively charged PEC particles composed of PDADMAC and PMA-MS (n–/n+=0.6, $c_{Pol,\,tot}$=0.0032 m) on a negatively charged surface

PDADMAC and poly(maleic acid-*co*-propylene) (PMA-P) were found to be more deformed and disk-like.

From this it was concluded that PEC nanoparticles can be somewhat analogous to latex systems, lacking, however, in an equivalent stability at the respective higher solid contents. The possibility of surface charge modification of inorganic negatively charged substrate materials is given by the choice of the PEC [96]. PECs have also been shown to be useful for flocculation processes [97].

2.6
Polyelectrolyte-Multilayers

The most common procedure for the production of polyelectrolyte multilayers (PEMs) at solid substrates is the layer-by-layer technique [3]. Cationic and anionic polyelectrolytes are deposited alternately from aqueous solutions to a target surface. The method is extremely versatile because, in addition to polyelectrolytes, numerous charged nanosized objects such as molecule aggregates, clusters, proteins and colloids can be deposited as multilayers. The internal structures of these multilayers is a target area of interest for working groups using or being involved in the PEM concept.

A model was proposed by Ladam and Decher [98], whereby the multilayer phase is divided into 3 zones : The inner core zone I close to the substrate consists of few (4-5) inhomogeneously adsorbed PEL layers and zone II is a homogeneous isotropic layer phase, in which a 1:1-stoichometry and high entanglement prevails. The outermost zone III consists again of a few (4-5)

PEL layers whose charges are not completely compensated. The uppermost PEL layer it is claimedis released into zone II, when a new oppositely charged polyelectrolyte is adsorbed on top. The isotropic unordered internal structure of zone II might be described theoretically by the Random-Phase-Approximation (RPA) [99], according to which the complexed and entangled PELs are modeled by Gaussian coils. This model was supported by Castelnovo and Joanny [100], who gave a mechanism for the overcharging as the driving force for the multilayer built up.

Generally, Zone (II) of PEMs is seen to be identical with the inner zone of dispersed polyelectrolyte complexes (PECs, see above), whereas the zone III of PEMs might be analogous to the outer shell zone of PECs.

2.6.1
Dissociation degree of PEMs and PECs

In Fig. 7 IR spectra of PEM-5 and PEM-6 of PEI/PAC and of solution cast PECs of PEI/PAC (n−/n+=0.80 and 1.25) prepared at pH=4 (PAC) and 6 (PEI) are shown. Both the spectrum of the polycation terminated PEM-5 and that of PEC-0.8 of PEI/PAC were comparable, since they contain PEI in excess. The PEM-5 spectrum showed only a minor $v(C=O)$ and a major $v(COO^-)$ band, since all COOH groups were consumed by the overlaying PEI layer, whereas the PEM-6 spectrum showed a major $v(C=O)$ and minor $v(COO^-)$ band. From these two bands the dissociation degree of PAC can be calculated using the relation

$$\alpha_{IR} = A_{v(COO^-)}/(1.74 \cdot A_{v(C=O)} + A_{v(COO^-)}) \tag{2}$$

which was described in [93]. Using this formula, the α_{IR} values of differently deposited PAC containing multilayers or PECs can be calculated, which will

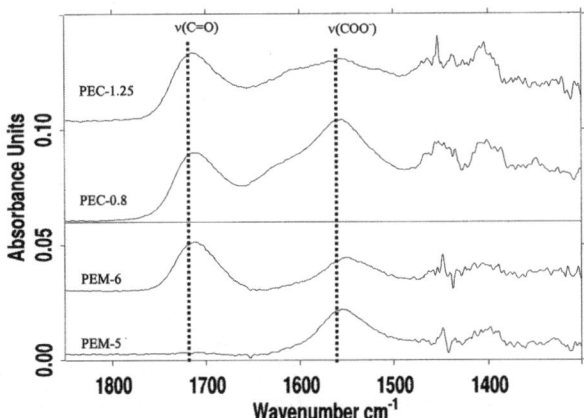

Fig. 7 ATR-FTIR spectra of the consecutively adsorbed PEM-5 and PEM-6 and of solution casted PECs of PEI/PAC with n−/n+=0.80 (PEC-0.8) and 1.25 (PEC-1.25)

Table 1 Dissociation degree α_{IR} of several samples of PEMs and PECs composed of PAC and PEI or PDADMAC deposited or mixed at pH 6 (PEI) and 4 (PAC)

	α_{IR} PEM-5	α_{IR} PEM-6	α_{IR} PEC-0.80	α_{IR} PEC-1.25
PEI/PAC	0.89±0.03	0.26±0.03	0.40±0.05	0.20±0.05
PDADMAC/PAC	0.70±0.03	0.13±0.03	low	low

be reported more detailed in [101]. In the Table 1 α_{IR} values are given for the samples introduced in Fig. 7. Significantly, for the polycation terminating PEM-5 of PEI/PAC the dissociation degree gets close to unity, which means all COOH groups are deprotonated by the reacting ammonium groups of the polycation. Whereas for the PAC terminating PEM-6 the dissociation degree was $\alpha_{IR} \leq 0.3$. From this we can learn that nearly all underlying COOH groups of the PAC were consumed by PEI and a newly adsorbed PAC provides again for new COOH groups to be dissociated by the next PEI layer.

For the PEM-5 of PDADMAC/PAC, not all COOH groups of the underlying PAC layer were consumed by PDADMAC (α_{IR}=0.70) and α_{IR} was significantly lower in the PEM-6 (α_{IR}=0.13). From this we can learn that in PDADMAC/PAC multilayers, fewer COOH groups are consumed than in PEI/PAC multilayers. This may be explained by the higher flexibility of PEI and the higher stiffness of PDADMAC. Since PAC at pH 4 should be in a coiled state, the stiffer PDADMAC might be not able to match all the carboxylic acid groups of PAC in a sufficient way, which is possible to a higher extent for the flexible PEI. PEMs of PDADMAC/PAC may therefore deviate more from 1:1 charge stoichiometry having higher uncompensated loop contributions, whereas PEMs of PEI/PAC are closer to 1:1 stoichiometry having fewer loops.

Similar relative results were found for the polyelectrolyte complexes: the α_{IR} of PEC-0.8 of PEI/PAC (0.4) was significantly higher than that of PEC-0.8 of PDADMAC/PAC. Also, according to Table 1, the α_{IR} of the PEC-0.80 was higher than that of the PEC-1.25 (0.20), which might be qualitatively analogous to the higher α_{IR} of the polycation terminated PEM-5 compared to PEM-6, respectively.

2.6.2
Multilayers of PECs

As a further surface modification, concept PEMs may be deposited using negatively and positively charged PECs. Recently, the consecutive adsorption of PEC$^+$ (n−/n+=0.6) of PDADMAC/PMA-MS with the respective PEC$^-$ (n−/n+=1.6) was reported [94]. In contrast to oppositely charged polyelectrolytes, a lateral growth of the PECs was obtained, which was evident from ATR-FTIR spectral data. Obviously, the oppositely charged PEC nanoparticles were growing laterally at the surface in contrast to the vertically growth of polyanions and polycations on top of eachother in PEMs from the 5th layer on, in which a homogeneous surface layer is established. These data illustrate that using spherical oppositely charged PECs, homogeneous polyelectrolyte layers can be built up. As it is previously shown (Fig. 6) this is not the case for uniquely charged ones (either PEC^+ or PEC^-) on oppositely charged surfaces, since there is a general jamming limit of adsorbing particles of about 55% as it is reported in [102]. However this surface modification concept might actually be too sophisticated at the moment for industrial applications and has to be simplified by experiment.

Furthermore, negatively charged PEC particles (PEC-1.6) of PLL and PMA-P can be consecutively adsorbed with PEI as the polycation to give homogeneous surface coverages, which is reported in [95].

2.6.2.1
Anisotropic Multilayers

A major challenge today is how to improve the orientation and the stability of the multilayers and how to characterize them properly.

Laschewsky et al., for example, formed organic-inorganic hybrid multilayers of polyelectrolytes and a clay that were stabilized by photocrosslinking [103]. Müller deposited cationic poly(L-lysine) (PLL) and polyanions like poly(maleic acid-co-α-methylstyrene) (PMA-MS) or poly(vinylsulfate) (PVS) by consecutively adsorbing on unidirectionally scratched silicon substrates [104]. The poly(L-lysine) formed α-helical rods in the multilayers that are aligned along the scratching direction of the substrate. He applied attenuated total reflection-Fourier transform spectroscopy (ATR-FTIR) using polarized light for the determination of the orientation within the films.

Representative dichroic ATR-FTIR spectra of consecutively adsorbed multilayers of PLL/PMA-MS in contact with an α-helix inducing 1m $NaClO_4$-solution are given in Fig. 8 for the untexturized substrate (A) and the texturized one (B). A significant variation of the relative amide I and amide II peak heights in the p and s-polarized ATR-FTIR spectra can be observed for the texturized substrate. From the dichroic ratios R of the amide I and amide II peak heights taken from the p-polarized and s-polarized spectra (i.e. $R_{amide\ I}=A_p(amide\ I)/A_s(amide\ I)$ and $R_{amide\ II}=A_p(amide\ II)/A_s(amide\ II)$), respectively, an order parameter S could be determined as it is described in [104]. Order parameters for α-helical PLL were found ranging from S=0.4 to 0.7 (S is equal to 1 for perfect orientation) in dependence of the chosen polyanion and parameters like the number of adsorption steps and the hydration state (dry/wet) of the sample. The degree of texturization and the molecular weight of PLL were found to be of main importance for the uniaxial alignment of the complexed α-helical rods [105]. Surfaces of anisotropically oriented α-helical polypeptides are interesting for biomimetic purposes. Further, oriented optically active rigid-rod polyelectrolytes [106] can become an essential building-block in nanosized electronic devices [107]. ATR-FTIR spectroscopy has proved itself to be very useful for a quantitative

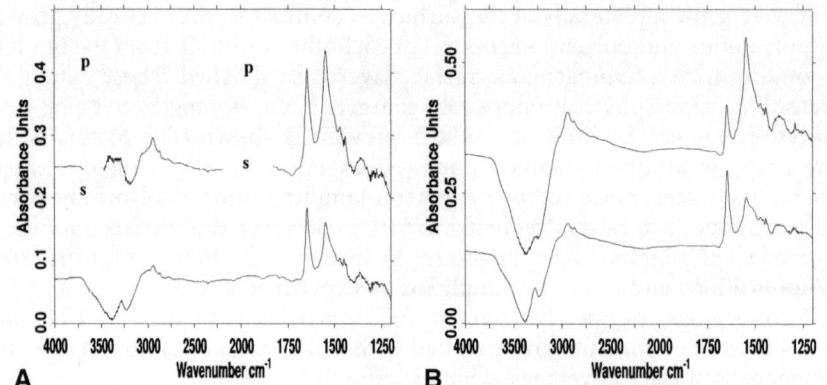

Fig. 8 *p*- and *s*-polarized in-situ-ATR-FTIR spectra of PLL/PMA-MS multilayers recorded after 10 deposition cycles on the untexturized (**A**) and on the uniaxially scratched Si substrate (**B**) in the presence of 1m NaClO$_4$ (wet state) in the range 4000-1200 cm^{-1}

in situ detection of the multilayer deposition and molecular changes within the multilayer phase induced by external stimulation [93]. Well investigated stimuli are the changing the pH [108], organic solvents [109], and humid air [93].

Concerning the latter, PEM composed of PEI and PMA-MS were described, which were deposited on silicon substrates and the water uptake from humid air (90% rel. hum.) of the dry sample was measured by ATR-IR spectroscopy. In Fig. 9, band areas of the ν(OH) band, which were normal-

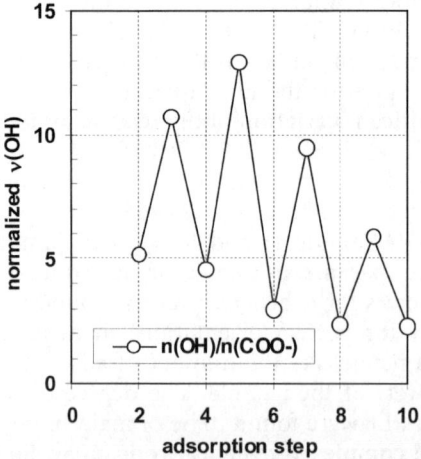

Fig. 9 Water uptake of the PEM-PEI/PMA-MS at 90% r.h in dependence of the adsorption step. The ν(OH) band area of difference spectra between hydrated and dry films was normalized by the ν(COO$^-$) at 1580 cm^{-1} (PMA-MS) to obtain a measure for thickness independent water uptake of the PEM

ized by the $\nu(COO^-)$ of the PMA-MS, are shown. By this normalization, the effect is compensated in that thicker layers absorb more water, so that the plotted data are a direct measure of thickness-independent water uptake. A modulated course of the water uptake can be observed in dependence on the adsorption step, whereby all polycation-terminated PEMs show a higher water uptake than that of the polyanion terminated ones. Moreover, wetting measurements (sessile drop) revealed a lower contact angle for the PEI-terminated PEM (43°) compared to the PMA-MS-terminated one (67°).

Similar results to these were found by Schwarz and Schönhoff [110] on nanoparticles coated by the PEM composed of poly(allylamine) (PAA) and PSS via ^{1}H NMR relaxation measurements. There, an increased water immobilization is reported for all odd polycation (PAA) terminating PEMs in contrast to even polyanion (PSS) terminating PEMs, which was found to be not just a property of the last adsorbed layer but more of the whole multilayer phase.

2.6.2.2
Protein/PEM Interaction

The possibility of a selective interaction between proteins and the outermost surface of polyelectrolyte multilayers [108] may become of outstanding importance in the development of engineering protein resistant surfaces for various biomedical applications by wet chemical surface modification.

As an example, results on the adsorption of human serum albumin (HSA) at a hydrophobic polypropylene (PP) film, which was modified by PEI/PAC multilayers are shown in Fig. 10 [111]. Based on ATR-FTIR spectra in the amide I region of the adsorbed protein layer the medium kinetic course

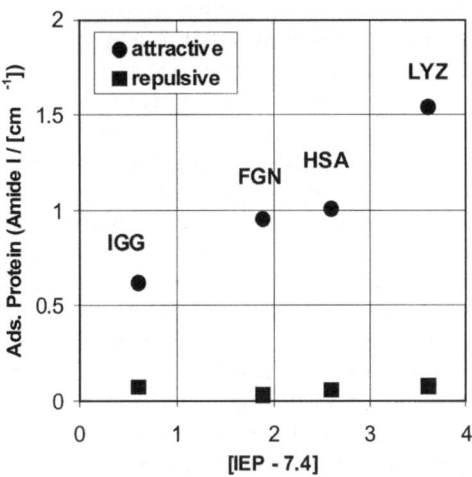

Fig. 10 Adsorbed amounts of the proteins HSA, FGN, IGG and LYZ at PAC (4) or PEI (5) terminating PEI/PAC multilayers in dependence of their IEP

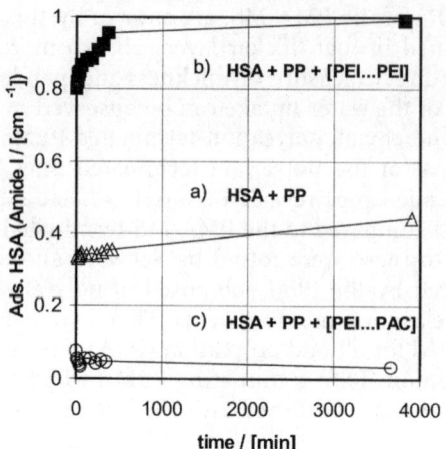

Fig. 11 Adsorbed HSA amounts on an unmodified and multilayer modified polypropylene (PP) film: (*a*) bare PP film, (*b*) CO_2-plasma treated PP-Film, modified by PEI (5) terminating PEM of PEI/PAC, (*c*) CO_2-plasma treated PP-Film, modified by PAC (4) terminating PEM

shows the moderate adsorption of HSA at the bare PP film, the lower the small adsorbed amount at the PP film modified by the polyanion capping multilayer and the upper course at the high adsorbed amount at the polycation terminating layer, respectively. Hence it is possible to modify hydrophobic polymer substrates by the PEM concept using commercial polyelectrolytes like PEI/PAC in order to control the protein affinity mainly by electrostatic interactions. In that framework especially weak polyanions are obviously very suitable to improve the protein resistance of surfaces towards negatively charged blood plasma proteins. As previously reported in [108], strong polyanions such as poly(vinylsulfate), whose linear charge density is not pH dependent, are not so repulsive as weak ones such as PAC, which are pH dependent. This might be explained by proteins being composed of amino acids with weakly charged groups because proteins do not contain amino acids with strong acids in their side chain groups. Therefore the carboxylate (Asp, Glu) or ammonium groups (Lys) containing amino acids of proteins are influenced in approximately the same way as the weak polyelectrolytes by the pH of the medium causing electrostatic attraction or repulsion.

An example how differently charged proteins are selectively adsorbed by PEI/PAC multilayers is given in Fig. 11. There, the adsorbed protein amount is plotted against the magnitude of the difference between the solution pH and the isoelectric point of the respective proteins (HSA, IGG, LYZ, FGN) [108] for both the electrostatically repulsive and the attractive case. Obviously the attractive interaction between the outermost surface of PEI/PAC multilayers and oppositely charged proteins scaled with the protein IEPs in contrast to the repulsive protein interaction, which showed no such dependence.

3
Polyelectrolyte-Surfactant Complexes (PE-surfs)

The complexation of polyelectrolyte with surfactants, which include numerous low and medium weight amphiphiles as well as lipids, has been investigated for many years nearly exclusively in solution as reviewed by Goddard [112]. Since 1994, starting with work by Antonietti et al. [15], polyelectrolyte-surfactant complexes (PE-surfs) in the solid state have also become of increasing interest. For some reviews see [1, 4-7, 113].

3.1
PE-Surfs in the Solid State

The most characteristic feature of solid-state PE-surfs is that they form a large number of mesomorphous structures which are liquid crystalline-like. These range from flat lamellar structures [114, 115] and discotic columnar structures [116] to periodic structures-*within*-structures [117]. The latter were first described by ten Brinke et al. [118]. The major factor that determines the type of the mesophase is the surfactant while the polyelectrolyte acts as a kind of glue between the charged surfactant head groups. Therefore, the polymer has normally the major influence on the mechanical properties of PE-surfs and not on the type of the mesophase [119]. But this is only a general rule. Chu et al. have shown that the charge density of the polyelectrolyte can also have a significant structural effect [120]. It has further been shown that the number of perforations increases in a perforated lamellar PE-surf system when the charge density of the polyelectrolyte increases [121].

In addition to the tuning of the structures, much attention is currently being paid to the physical properties. Tieke, for example, has produced mesomorphous photoluminescent PE-surfs [122] and we have reported PE-surfs that display electroluminescence [123]. In one case it was found that the electroluminescence spectrum can be tuned simply by varying the surfactant [124]. Faul et al. have shown recently that optically functionalized complexes can form highly organized nanostructures even when the polyelectrolyte is substituted by dyes that contain three or four charges [125, 126]. They show that the binding of surfactant molecules to a polycharged dye is a cooperative process as is known from typical PE-surfs. This proves that the concept of forming highly ordered mesomorphous structures from PE-surfs by a cooperative binding process can be transferred to dye-surfactant complexes. Similarly, GSSG, a charged oligopeptide, was complexed with surfactants to form well-ordered structures [127].

3.2
Dispersions and Nanoparticles

Three different ways have been developed to produce nanoparticle of PE-surfs. The most simple one is the mixing of polyelectrolytes and surfactants in non-stoichiometric quantities. An example for this is the complexation of poly(ethylene imine) with dodecanoic acid (PEI-C12) . It forms a solid-state complex that is water-insoluble when the number of complexable amino functions is equal to the number of carboxylic acid groups [128]. Its structure is smectic A-like. The same complex forms nanoparticles when the polymer is used in an excess of 50% [129]. The particles exhibit hydrodynamic diameters in the range of 80-150 nm, which depend on the preparation conditions, i.e., the particle formation is kinetically controlled. Each particle consists of a relatively compact core surrounded by a diffuse corona. PEI-C12 forms the core, while non-complexed PEI acts as a cationic-active dispersing agent. It was found that the nanoparticles show high zeta potentials (approximate to +40 mV) and are stable in NaCl solutions at concentrations of up to 0.3 mol l^{-1}. The stabilization of the nanoparticles results from a combination of ionic and steric contributions. A variation of the pH value was used to activate the dissolution of the particles.

The second way of forming nanoparticles of PE-surfs is to use water-soluble block-copolymers with one complexable (ionic) block and one non-complexable (nonionic) block. For example, the mesomorphous complexes (solid state) between poly(ethylene oxide)--b-poly(ethylene imine)s and dodecanoic acid can easily by dispersed in water to form nanoparticles [117]. The nanoparticles are of core-shell type and have sizes around 200 nm. Their cores are formed by poly(ethylene imine) dodecanoate while their shells consist of poly(ethylene oxide). It was found that the shapes of the nanoparticles depend on the PEI block. They are, for example, prolate if the PEI is linear and spherical if the PEI is branched.

A variation of this approach towards redispersible PE-surf nanoparticles was presented by Hentze et al. [130]. Their particle formation procedure is as follows: in a first step they prepared core-shell particles with a weakly cross-linked core of poly(styrene) and a shell of poly(ethylene oxide). Secondly, the cores were sulfonated to poly(styrene sulfonate). The cores were complexed with cationic surfactants in the third and final step. They became mesomorphous due to the complexation and show a characteristic length that varies between 2 and 4 nm depending on the alkyl chain length of the surfactant. The mean size of the whole particle is around 400 nm.

The third way to PE-surf nanoparticles is based on the use of polyampholytes, i.e. polyelectrolytes that contain cationic and anionic groups. Examples are long-term stable fluorinated nanoparticles, prepared by the complexation of polyampholytes with perfluorododecanoic acid [131]. The polyampholytes are statistic copolymers of styrylmethyl(trimethyl)ammonium chloride, methacrylic acid, and methyl methacrylate having more cationic monomers (28-100 mol %) than anionic monomers (0-16%). The cationic monomers form a complex with perfluorododecanoate ions to create neutral entities that aggregate while the negatively charged methacrylic monomers prevent a macroscopic precipitation such as normally observed in fluorinated complexes. Negatively charged nanoparticles, which have zeta potentials of about −40 mV, are formed as a result of the combined complexation and stabilization. Small-angle X-ray scattering investigations reveal that the particles are anisotropic. For example, disk-shaped particles with a diameter of D greater than or equal to 30 nm and a height of 2.2 nm were formed for a high number of complexing groups (79 mol %). Cylindrical-shaped particles with lengths of H greater than or equal to 25 nm and radii of 1.5 nm were produced for a medium number of complexing groups (57 mol %). Fluorescence spectroscopy with pyrene as the probe was used to determine the critical aggregation concentrations of the particles which are in the range 0.01-0.06 g

The smallest known particles are prepared by using polyampholytes, that have short chain lengths, and with alternating cationic and anionic monomers along the polymer chain [132]. Fatty acids (dodecanoic acid and perfluorododecanoic acid) were used for complexation. The formation of the polyelectrolyte-fatty acid complexes is self-assembled and generates nanoparticles with sizes in the range of 3-5 nm that are named dressed micelles. A defined arrangement of the ionic charges of three polyampholytes was achieved by the copolymerization of a cationic vinyl monomer (N,N'-diallyl-$N''N$-dimethylammonium chloride) and anionic vinyl monomers (maleamic acid, phenylmaleamic acid, and 4-butylphenylmaleamic acid). The zeta potentials of the dressed micelles were adjusted in the range of −56 to 25 mV. They increase when replacing the alkylated dodecanoic acid by its perfluorinated counterpart, and they also increase when enhancing the hydrophilicity of the polyampholyte.

3.3
Drug Carriers

In addition to a better basic understanding of self-assembly processes one of the most attractive aims of the preparation nanoparticles of PE-surfs is the development of new drug delivery systems. These can be either simple drug carrier systems-or even more challenging-drug targeting systems. Promising candidates for drug carriers are complexes of poly(ethylene imine)s and fatty acids. They show loading capacities of hydrophobic drugs (e.g.,triiodothyronine and Q_{10}) in the range of 15 to 20% (w/w) [129].

Polyelectrolyte complexes of retinoic acid have been well investigated, they are pharmaceutically active surfactants. In the following, we will therefore discuss the physicochemical properties of drug carriers formed by synthetic polyamino acids, polyethyleneimine, double hydrophilic block copolymers and retinoic acid.

Vitamin A and its analogues, in particular retinoic acid, are involved in the proliferation and differentiation of epithelial tissues and have continued to be used in the treatment of dermatological disorders such as acne, psoriasis and hyperkeratosis [133, 134]. Currently, much effort is being focused on how to understand the role of retinoic acid in cell differentiation. This is done by investigating the binding properties of retinoids on specific proteins [135] Their role in malignant-tumor inhibition [136, 137] and in their regulation of brain functions [138]. Natural retinoids need to be bound to specific retinoic-binding proteins in order to ensure their protection, solubility, and transport by body fluids. Immobilization is a major problem in administering retinoic acid as a pharmacological agent. One way of achieving such immobilization and protection of retinoic acid is by binding it to a protein as is the case in nature. A successful example for mimicking nature's strategy was shown by Zanotti et al. [139] when he cocrystallized transthyretin and retinoic acid. This, however, is a difficult and costly procedure. An easier and less expensive method for the required immobilization by the complexation of retinoic acid with synthetic cationic polyelectrolytes such as poly(ionene-6,3), poly(N-methyl-4-vinyl-pyridinium) and poly(diallyldimethylammonium) has been developed [140, 141].

The work presented in the next section concerns the immobilization of retinoic acid by three polyamino acids: poly(-L-lysine), poly(-L-arginine) and poly(-L-histidine). The mesomorphous structure of these complexes, which are prepared as nano-particles, has been examined. Then we will report on the physicochemical characteristics of water-soluble complexes formed between PEO-PLL block copolymers and retinoic acid. The structure in the solid-state and solution are discussed. Further the dissociation of the complex when changing the pH is reported. Finally, the immobilization of retinoic acid by PEI with different molecular weights is presented.

3.3.1
Immobilization of Retinoic Acid by Polyamino Acids [142]

The retinaote complexes of poly(L-arginine) (PLA), poly(L-histidine) (PLH) and poly(L-lysine) (PLL) were precipitated from aqueous solutions, purified and cast from solution in 2-butanol in order to form films. A drawing of the compounds used is shown in Fig. 12. In comparison to complexes of retinoic acid with other poly cations, such as poly(ionene-6,3), poly(N-methylene-4-vinylene-pyridinium) and poly(diallyldimethylammonium) [140], the dissolution rate of the polyamino acid complexes was significantly slower. In addition, films prepared from the former are flexible, have glass-transition temperatures in the range of -19 to 28 °C and show viscoelastic mechanical properties, whereas those of PLA, PLH and PLL are brittle. Differential scanning calorimetry showed no glass-transitions. The difference in the solubility and mechanical properties of retinoate complexes of polycations containing an atactic backbone and complexes with basic poly(L-amino) acids may be explained by the stereo-chemically unique polymer backbones of the latter, which enhance the stiffness of the retinoate complex. It has already been stated that no glass-transitions were found in solid-state complexes of poly(L-lysine) with alkyl sulfates [143] or soylecithin [144]. Therefore, at least for poly(L-lysine), the absence of a glass transition may be assumed to be a characteristic property of its solid-state complexes with surfactants.

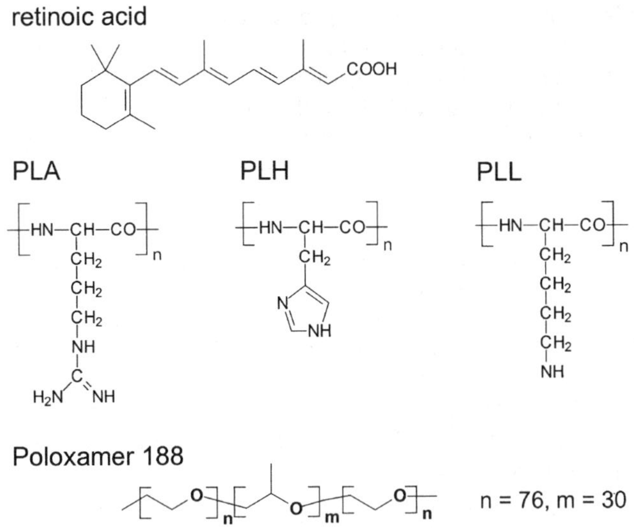

Fig. 12 Substances used for complex formation and nanoparticles. (retinoic acid) All-*trans* retinoic acid, (PLA) poly(-L-arginine), (PLH) poly(-L-histidine), (PLL) poly(-L-lysine). As dispersing agent a tri-block copolymer was used. It consisted of ethylene oxide and propylene oxide (Poloxamer 188). Reprinted with permission from [142]. Copyright 2000 American Chemical Society

3.3.1.1
Chain Conformation

The conformational state of the polymeric backbone of the complexes was examined by circular dichroism (CD). Figure 13 shows the CD spectra of

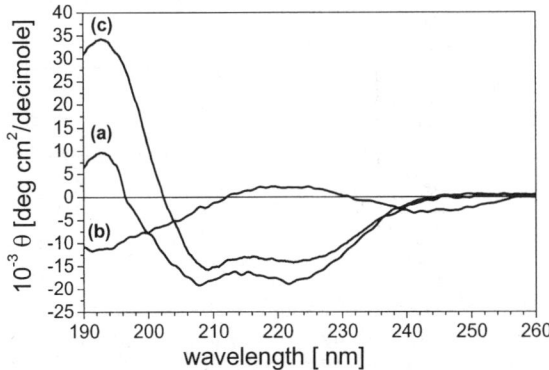

Fig. 13 Circular dichroism spectra of poly(L-arginine) retinoate (*a*), poly(L-histidine) retinoate (*b*), and poly(L-lysine) retinoate (*c*). The complexes were prepared as films on quartz slides. Reprinted with permission from [142]. Copyright 2000 American Chemical Society

complex films prepared on quartz slides. A maximum at 191 nm and two minima at 210 and 222 nm were found for poly(L-arginine) retinoate (curve a in Fig. 13) and poly(L-lysine) retinoate (curve c in Fig. 13). The spectra of complex solutions in 2-butanol show the same characteristics. Therefore, it was concluded that the conformation of the polymeric chains of both complexes in the solid state and in a solution of 2-butanol is an α-helix [145, 146]. By contrast, a minimum at 193 nm and a maximum at 222 nm was observed in the CD spectrum of poly(L-histidine) retinoate (curve b in Fig. 13). This proves that there is a predominantly coil conformation for poly(L-histidine) retinoate. Since the pK_a of the imidazole group of histidine residues is around 6.6 [147], the uncharged forms of histidine were present at the complex preparation condition (pH 9). By contrast the poly(L-arginine) and poly(L-Lysine) are charged at pH 9. The pK_a-values of the basic groups are 12.48 and 10.53, respectively [148]. On the basis of this data, it may be assumed that the charging of the polyamino acid leads to an α-helix conformation of the complex.

3.3.1.2
Solid-State Structures

Information about the molecular packing of the retinoate moieties was obtained by wide-angle X-ray scattering (WAXS). The WAXS pattern of retinoic acid (curve a in Fig. 14) is characterized by a number of sharp reflections resulting from the high degree of crystallinity. Retinoic acid crystallizes in two similar crystalline modifications (triclinic and monoclinic) [149], which produce the reflex pattern observed. As shown in Fig. 14, curves b and d,

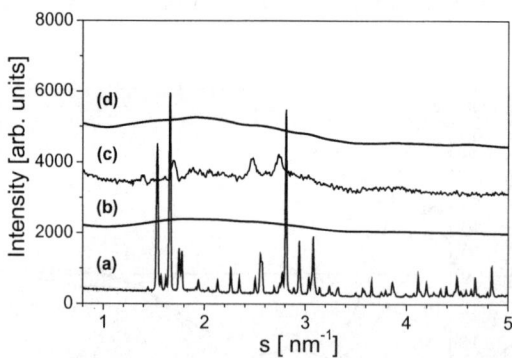

Fig. 14 Wide-angle X-ray diagrams of crystalline all-*trans* retinoic acid (curve *a*) and its complexes with polyamino acids. In the curve of poly(L-arginine) retinoate and poly(L-lysine) retinoate (curves *b* and *d*, respectively) no crystalline reflections were found. However, three weak crystalline reflections are present in the curve of poly(L-histidine) retinoate at s=1.69, 2.48 and 2.73 nm^{-1} (curve *c*). This is indicative of a partially crystalline complex structure. Reprinted with permission from [142]. Copyright 2000 American Chemical Society

the crystallinity of retinoic acid was prevented by complexation with poly(L-arginine) and poly(L-lysine). As is the case in complexes with polydiallylamonium chloride, ionene-3,6 and poly(N-methyl-4-vinylpyridinium chloride) [140], the maximum of the broad wide-angle reflection corresponds to a Bragg-spacing of 0.52 nm. The absence of sharp reflections in the wide-angle region proves the lack of crystallinity. Again, the properties of poly(L-histidine) retinoate differ from that of the others. Weak but significant reflections at $s=1.69$, 2.48 and 2.73 nm^{-1} were observed which are indicative for the partial crystallinity of the complex (curve c in Fig. 14). The scattering patterns remained constant for several months, and therefore it was assumed that the complexes are likely to be thermodynamically stable. The absence of crystallinity, is an interesting aspect, both from a material science as well as from a pharmaceutical point of view, because the weight percentage of the crystallizable retinoic molecules in all complexes is about 70%. The strong reduction of crystallinity was explained as follows: In order to maintain electrostatic neutrality, any diffusion of retinoic moieties must be accompanied by a correlated movement of charged polymer chains, such that no phase separation, and consequently no crystallization, can occur.

3.3.1.3
Nanoparticles

It was expected that the retinoate complexes of PLA, PLH and PLL form lamellar nanostructures similar to those reported earlier [140].The films of all complexes are optically anisotropic, as found during examination between crossed polarizers. Obviously the complexes are mesomorphous, but an unambiguous identification of the mesophases on the bases of the optical properties was not possible. Therefore, the state of order on a length scale of several nanometers was investigated by small-angle X-ray scattering measurements carried out on freeze-dried complex nano-dispersions. Aqueous nano-dispersions were prepared from powder of the complex with the aid of poloxamer 188, a triblock copolymer consisting of two polyethyleneoxide blocks and a polypropyleneoxide (see Fig. 12). Poloxamer 188 is a common non-ionic surfactant frequently used as a steric stabilizer in pharmaceuticals and already proved to be capable of stabilizing nanoparticles [150-152].

The size of the complex particles as determined by dynamic light scattering was in the range of 300 to 380 nm in diameter with a polydispersity of about 0.20. It was found that the complex dispersions could be redispersed after freeze-drying. As a result of this procedure the particle sizes increases slightly to about 400 nm and the polydispersity increases to values of about 0.3. The simplicity of the preparation of these nanoparticles seems to be an attractive feature of these complexes. As in these complexes, in the small-angle X-ray scattering diagrams of nano-dispersions of retinoate complexes equidistant reflections are also present (see Fig. 15). It can be seen that the positions and the sharpness of the scattering peaks are different for the complexes. In the case of poly(L-arginine) retinoate, a lamellar repeat unit of 3.62 nm was determined by fitting a Lorentzian peak profile onto the (001)

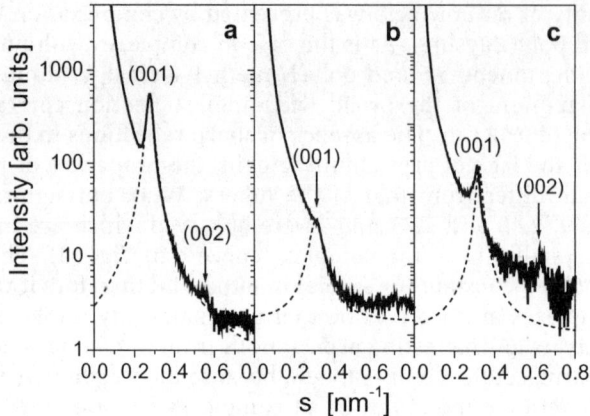

Fig. 15 Small-angle X-ray scattering curves of poly(L-arginine) retinoate (*a*), poly(L-histidine) retinoate (*b*) and poly(L-lysine) retinoate (*c*). The dashed lines are fits using a Lorentzian peak profile. Repeat units are 3.62 nm, 3.27 nm, and 3.10 nm. Reprinted with permission from [142]. Copyright 2000 American Chemical Society

reflection. The corresponding values for poly(L-histidine) and poly(L-lysine) are 3.27 nm and 3.10 nm, respectively. Obviously the size of the repeat unit increases with an increasing mass of the monomer. In addition, the increase of the size of the repeat unit is linear with respect to the mass of the monomer unit (+0.0185 nm per mass unit). This linear scaling is indicative of the smectic A-like structures of the three complexes.

The correlation lengths of the structures were determined from the reciprocal width of the reflections; they were 27±1, 15±3 and 38±1 nm (arginine, histidine and lysine). The correlation length values can be explained by an increasing stacking order of the smectic layers from poly(L-histidine) retinoate, poly(L-arginine) retinoate to poly(L-lysine) retinoate. Such differences may be the result of packing constraints due to their different molecular geometry, which may lead to different degrees of frustration [15]. The question arises as how to characterize the internal lamellar particle structure in detail. Parameters which are easily available are only the size of the particles and the size of the repeat units within them. In earlier works on polyelectrolyte surfactant complex films, it was found that a good model for the description of their mesophase structures could be made by the microphase separation of the ionic rich and hydrophobic rich regions. Often, the density transition between these regions is given by a step function, which is similar to that found for strongly segregated block copolymers and can be identified by the presence of Porod's law [182,183]. With such ideal phase-separated lamellar sheets within the complex nanoparticles, two limiting cases for their internal structures were assumed: an 'onion'-like and a 'tart'-like structure. As demonstrated earlier [142] it must be stated that currently the polydispersity (20%) of the complex nanoparticles is too large to produce de-

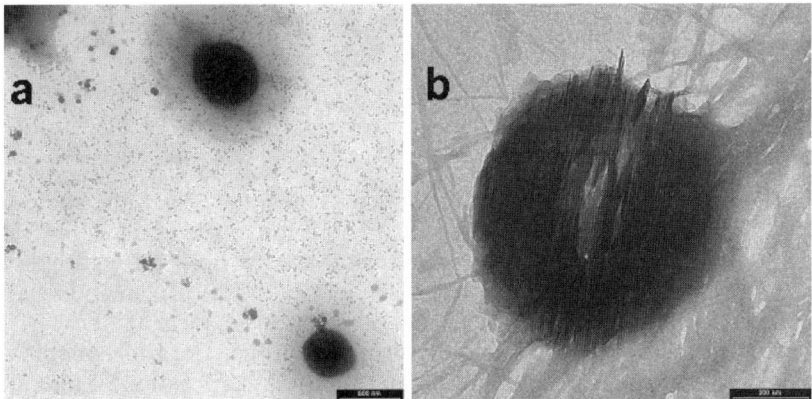

Fig. 16 Typical TEM pictures of a dried dispersion on a copper grid. (*a*) At lower resolution spherical complex particles are shown surrounded by a halo consisting of block copolymers (scale bar=500 nm). (*b*) At higher resolution, the internal structure of the particle is revealed to be tart-like (scale bar=200 nm). Reprinted with permission from [142]. Copyright 2000 American Chemical Society

tailed information about the nature of their internal structures such as onion- or tart-type by small-angle X-ray scattering.

For more insight into the internal particle structure, electron microscopy experiments were carried out. In TEM pictures of dried complex nano-particle dispersions (Fig. 16 a) it can be seen that the nanoparticles consist of a core with a high contrast, which is surrounded by a shell of lower contrast. It was believed that the core consists of the complex and the shell of the dispersing agent, Poloxamer 188. At a higher resolution (Fig. 16b) it can be seen that the internal structure of a dried particle is split into discrete layers, extending form one side of the particle to the other. This is very much indicative of a more tart-type particle structure. Therefore, on the basis of electron microscopy data it seems to be very likely that the kind of finite lamellar particle structure is much more of a tart-type than an onion-type [142]. As is the case in small-angle X-ray investigations currently being undertaken, it was not possible to determine the values of the lamellar thicknesses d_1 and d_2 by electron microscopy. The reason for the formation of the tart-type particles is not understood yet but may be due to their preparation from films. In films the lamellar domain size is likely to be much larger than the size of the nano-particles and tart-type.

In conclusion it was shown that the solid-state complexes formed by poly(L-arginine), poly(L-histidine) and poly(L-lysine) cations and retinoic acid can be prepared as films and nano-particles. The high content of retinoate moieties, the absence of crystallinity and low particle sizes could make these complexes interesting as a new carrier for the delivery of retinoic acid, either transdermal or in body fluids. It may be speculated that supramolecular structures such as the smectic A-like nano-particles of the tart-type presented here can lead to interesting release profiles of pharmaceutically active agents.

3.3.2
Block copolymers [153]

Micelles formed by amphiphilic block copolymers have been the subject of intense research during the last decade. In addition to their interesting physicochemical properties, they are promising as carriers for hydrophobic drugs. A recent review was given by Eisenberg et al. [154] In an aqueous environment, the hydrophobic blocks of the copolymer form the core of the micelle while the hydrophilic blocks form the corona. The hydrophobic micelle core serves as a microenvironment for the incorporation of lipophilic drugs, while the corona serves as a stabilizing interface between the hydrophobic core and the external medium. It was shown that complexation of poly(ethylene oxide)-b-poly(L-lysine)s (PEO-PLL) with DNA as an oppositely charged polyelectrolyte leads to the formation of water-soluble complexes in an aqueous environment [155]. PEO-PLL block copolymer micelles were used to explore the feasibility of polymeric micelles as a novel vector system for genes and oligonucleotides [156, 157]. Supramolecular association of a block copolymer consisting of PEO and polyamino acids through hydrophobic or electrostatic interaction leads to the formation of core-shell type micelles in which drug molecules are hydrophobically or electrostatically included in the core of the particle which is surrounded by the PEO outer shell [158]. The proper micelle size of about 50 nm and the hydrophilicity of the outer-shell PEO seem to contribute to an extension of the period for which the drug circulates in the blood. Exceptionally high accumulations in a solid tumor were demonstrated for copolymer micelles with an entrapped anticancer drug (doxorubicin) [159, 160]. Kabanov et al. proposed a similar system, the complex of poly(ethylene oxide)-g-polyethyleneimine and biological active surfactants, as a novel drug delivery system [161]. Such complexes form 'micellar microcontainers' with retinoic acid, a molecule which is highly optically active and is considered to be a surfactant from the physicochemical point of view. In the previous chapter it was shown that nanoparticles of complexes of synthetic polyamino acids with retinoic acid contain an internal smectic A-like structure.

This section reports on the physicochemical characteristics of water-soluble complexes formed between PEO-PLL block copolymers and retinoic acid. Two block copolymers were used. The lengths of the PEO blocks in both are identical, the molecular weight was 5000 g/mol (M_w/M_n=1.1). This corresponds to a degree of polymerization of 114. The length of the PLL block varies: PEO-PLL18 has 18 L-lysine monomer units and PEO-PLL30 has 30. A sketch of the molecular structures is given in Fig. 17.

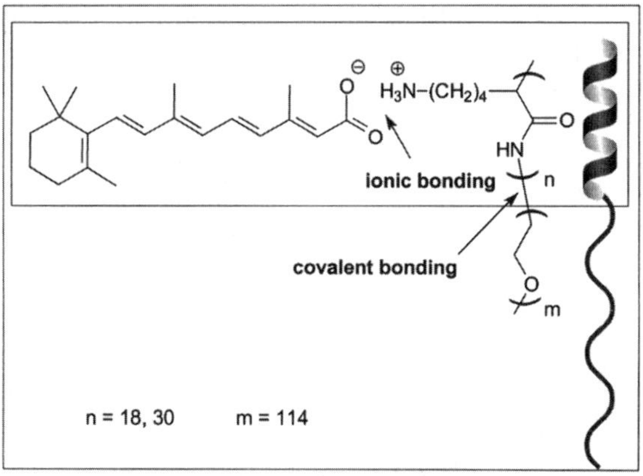

Fig. 17 The molecular structure of a poly(ethylene oxide)-*b*-poly(L-lysine) retinoate complex and a sketch of the polymeric backbone with an α-helical poly(L-lysine) block. Reprinted with permission from [153]. Copyright 2000 American Chemical Society

3.3.2.1
Crystallinity

DSC traces of the copolymers and of their complexes with retinoic acid in the solid state showed that the two non-complexed polymers undergo first-order transitions upon heating with a maximum at 56±1 °C. Upon cooling the transitions were found at 35±2 °C. The DSC traces of the complexes differ significantly from those of the polymers. A weaker transition is present in the curve of PEO-PLL18 retinoate at 47 °C when it is heated and at 23 °C when it is cooled. In the DSC curves of PEO-PLL30 retinoate, a very weak transition at 38 °C was detected on heating, while it was not found during cooling. No other transitions were observed in the temperature range of 0 to 150 °C. Homopolymers of poly(ethylene oxide) are known to form lamellar crystals with melt transitions that depend significantly on their molecular weights. As an example, the melting points of poly(ethylene oxide)s with narrow molecular weight distributions are found at 48, 58 and 64 °C for molecular weights of 1500, 3000 and 6000 g/mol respectively [162]. Earlier no melt transition had been found for the complex of a homopolymer of poly(L-lysine) and retinoic acid and the melting point of retinoic acid is 181 °C [163]. Therefore, the transitions in the copolymers and in their complexes were regarded as due to the melting of the crystalline poly(ethylene oxide) segments. Compared to a poly(ethylene oxide) homopolymer with a molecular weight of 5000 g/mol, its co-polymerization with a poly(L-lysine) block lowers the melting transition of poly(ethylene oxide) by about 5 °C. Here the melting point depression does not depend on the length of the lysine block. For the complexes the lowering of the melt transition temperature is much stronger and depends on the length of the lysine block (17 °C and 26 °C for PEO-PLL18 retinoate and PEO-PLL30 retinoate, respectively). The measured enthalpies of the melt transitions were compared with the perfect heat of fusion of PEO (203 J/g) [164] in order to estimate the degree of the bulk crystallinity. Taking into account the different amount of PEO in the compounds, the crystallinity of the PEO chains was found to be about 50% (PEO-PLL18), 45% (PEO-PLL30), 30% (PEO-PLL18 retinoate) and 10% (PEO-PLL30 retinoate). Compared to neat PEO whose crystallinity is in the range of 70 to 80% [164], the copolymerization with lysine reduces the degree of crystallinity by 20 to 30%. This is similar to the effect of blending PEO with PMMA [164]. In contrast to the copolymerization, the complexation reduces the degree of crystalline PEO even more strongly. Probably the complexed moieties disturb the formation of the extended chain conformation and the integral number of folded chain conformations, which are typical for the crystalline form of PEO [165]. The conclusion that crystalline PEO segments are the origin of the endothermic transitions was confirmed by wide-angle X-ray experiments. It was found that the X-ray scattering intensities of the PEO reflections at scattering vectors of $s=2.18$ nm^{-1} and 2.65 nm^{-1} decrease in the series PEO-PLL18, PEO-PLL30, PEO-PLL18 retinoate and PEO-PLL30 retinoate. This sequence of the intensities is in agreement with a decreasing amount of crystalline PEO segments in the samples derived by DSC.

3.3.2.2
Nanostructures in the Solid State

Small-angle X-ray scattering techniques were used to investigate the structure of the PEO-PLL polymers and their complexes on length scales in the range of 1 to 50 nm. The small-angle scattering diagrams of the PEO-PLL18 and PEO-PLL30 are essentially identical and show two Bragg peaks whose positions have a ratio of $1:3^{1/2}$ (not shown). This suggests a hexagonal arrangement of the poly(L-lysine) chains in a two-dimensional lattice with a center-to-center distance of 1.53 nm. This value is close to the chain spacing in poly(L-lysine) hydrochloride crystals (1.50 nm) [166]. In contrast to the pristine polymers, Bragg peaks with relative positions of 1:2 were found in the small-angle scattering diagrams of PEO-PLL18 retinoate and PEO-PLL30 retinoate, which indicate lamellar structures (see Fig. 18a,b). The long periods are 3.37 nm±0.05 nm (PEO-PLL18 retinoate) and 3.32 nm±0.05 nm (PEO-PLL30 retinoate), which is slightly larger than that found for a complex of a poly(L-lysine) homopolymer (M_v=15,000 to 30,000 g/mol) and retinoate, which is 3.10 nm [142]. This was attributed to a better packing of the latter as a result of the absence of packing constraints caused by PEO blocks. It can be seen in the scattering diagrams that the peaks of the PEO-PLL18 retinoate are broader than that of the PEO-PLL30 retinoate. The correlation lengths were 10 nm and 12 nm as determined from the width of the first order reflection. When taking into account the relative shortness of the lysine chains, the high degree of mesomorphous ordering is surprising.

Well defined conformations of the poly(L-lysine) chains have to be taken into account. It was found that in the solid state form of the non-complexed polymers, the poly(L-lysine) chains adopt mixtures of α-helix and β-sheet

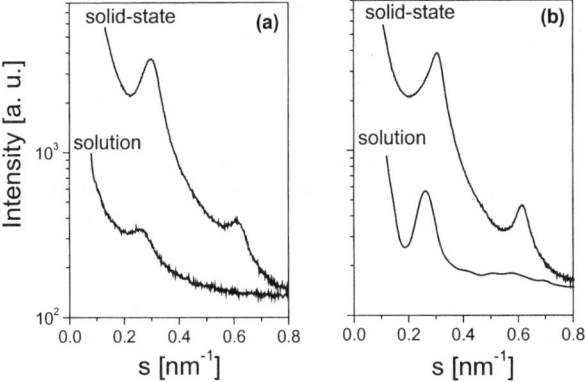

Fig. 18 Small-angle X-ray scattering curves of PEO-PLL18 retinoate (**a**) and PEO-PLL30 retinoate (**b**). Curves are given for the materials in the solid state (*upper curves*) and for dispersion in aqueous solution (*lower curves*). Reprinted with permission from [153]. Copyright 2000 American Chemical Society

conformations as shown by the position of the amide I and amide II vibrations in the FTIR spectrum [167] (see Fig. 19). It can be seen that the amide I vibrations of the α-helix is represented with a band at 1652 cm^{-1} and that of the β-sheet with a band at 1624 cm^{-1}. The individual amide II vibration bands overlap but the band at 1539 cm^{-1} is strongly indicative of significant amounts of β-sheet conformations. It was found that the intensity of the bands and consequently the amount of α-helix and β-sheet conformations depend to a significant degree on the preparation conditions such as solvent, temperature and drying time. Therefore it was not possible to identify differences in the conformations between PEO-PLL18 and PEO-PLL30. In contrast to the non-complexed polymers the positions of the amide I and amide

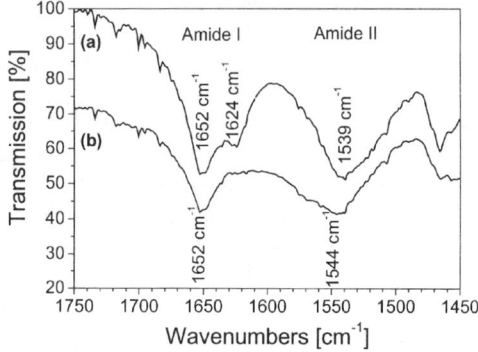

Fig. 19 FTIR spectra of the PEO-PLL18 (*a*) and the PEO-PLL30 retinoate (*b*) in the amide region. Reprinted with permission from [153]. Copyright 2000 American Chemical Society

II bands were found to be independent of the preparation conditions at 1652 cm^{-1} and 1544 cm^{-1} for PEO-PLL18 retinoate and PEO-PLL30 retinoate (Fig. 19b). This proves that the poly(L-lysine) chains strictly adopt an α-helix in the complexes. It was suggested that the α-helix in the PEO-PLL retinoate complexes is energetically more stable than the β-sheet conformation, which is probably due to a better spherical arrangement of the retinoate moieties when bound to an α-helix. Tirrell et al. reported on the structure of poly(L-lysine) complexes with alkyl sulfates which form lamellar structures [168]. They found that the α-helical and β-sheet portions of the poly(L-lysine) chains in their complexes depend to a great extent on the conditions of the preparation of the solid-state complexes from solution. The exact reason for the adjustment of the different chain conformation is not clear yet, but it was assumed that the formation of inter chain hydrogen bridges, which favors the β-sheet formation, is strongly suppressed for the two PEO-PLL retinoate complexes but not for the complexes of poly(-L-lysine) with alkyl sulfates.

3.3.2.3
Core-Shell Nanoparticles

Block copolymers containing a PEO block as a water-soluble segment and a second, water insoluble block, are known to be suitable for the formation of core-shell micelles, where the cores can be loaded with hydrophobic drugs [154]. Following this hydrophilic-hydrophobic concept the double hydrophilic PEO-PLL18 and PEO-PLL30 block copolymers were converted into hydrophilic-hydrophobic systems by their complexation with retinoic acid, which resulted in micellar solutions of the complexes. The solutions were investigated by X-ray scattering in an aqueous medium in the concentration range of 0.5 to 5% (w/w). The absence of reflections in the wide-angle scattering diagrams proved that no crystalline retinoic acid was formed during the complexation and that the PEO blocks are dissolved. In contrast to the wide-angle region sharp reflections were found in the small-angle diagrams of the micelles (see Fig. 15a,b) with Bragg spacings of 3.94 nm (PEO-PLL18 retinoate) and 3.83 nm (PEO-PLL30 retinoate). This was interpreted as resulting from mesomorphous core-shell micelles. The core which contain the PLL-retinoate moieties is of a lamellar structure and the shell is formed by PEO segments. The higher long period in the micelles compared to those of the complexes in the bulk material was explained by the incorporation of water in the core. From the differences of the long periods between the solid state complexes and the dispersed complexes the water content of the cores was estimated to be 17% (v/v) for PEO-PLL18 retinoate and 15% (v/v) for PEO-PLL30 retinoate. These values are close to the water uptake (18%) of a polyelectrolyte fluorosurfactant complex, in which the water is predominantly located around the polymeric backbone [169]. The location of the incorporated water and its mobilizing effect on the molecular dynamics is given in a detailed solid-state NMR investigation on polyelectrolyte-surfactant complexes [170].

From the widths of the reflections of the micellar solutions (Fig. 18) the correlation lengths of the mesophases within the cores were determined to be 6 nm (PEO-PLL18 retinoate) and 10 nm (PEO-PLL30 retinoate). It is known that, in a helical poly(amino acid), the amino acid monomer expands over 0.15 nm [171, 172]. In the case of poly(L-lysine)s (the polymerization degree of lysine is 18 and 30) the helical end-to-end distances were calculated to be 2.7 nm and 4.5 nm. Tentatively an α-helical conformation was assumed for the complexed PLL in the core of the micelles, the maximum lamellae stack size is about the same magnitude as the correlation lengths. Therefore it is probable that each micellar core contains a lamellar monodomain. Ten Brinke et al. have shown that a hierarchy of two different length scales (4.8 nm and 35.0 nm) can be produced when combining covalent bonding (block copolymer)s, proton transfer and hydrogen bonding [173]. PEO-PLL18 retinoate and PEO-PLL30 retinoate in the solid state could be considered to be similar to their block copolymer complexes. But in addition to the lamellar structuring of the PLL retinoate a regular structure with a larger long period (10 to 50 nm) could not be identified, which would have been expected from a microphase separation of PEO and PLL retinoate.

3.3.2.4
Helix-Coil Transition

A detailed investigation of the helix-coil transition of the non-complexed PEO-PLL18 has been reported on by Kataoka et. al [145]. By using circular dichroism they have shown that the PLL blocks of PEO-PLL18, which themselves cannot form an α-helix structure due to substantially lower molecular weight, form an α-helix structure in solution when they are bound to PEO. Their work shows clearly that the copolymerization with PEO has a stabilizing effect on the α-helical conformation of PLL. Bearing this in mind it was asked how the complexation of the PEO-PLL polymers affect the conformation of the PLL blocks. Circular dichroism was used as a sensitive method for the detection of the PLL chain conformations within the micelles. It was found that the shape of the circular dichroism spectra were neither dependent on the concentration of the complexes in a concentration range of 0.05 to 0.5% (w/w) nor does it depend on the storage time after preparation (1 h, 24 h, 72 h). In addition the spectra of PEO-PLL18 retinoate and PEO-PLL30 retinoate are identical within the range of experimental error. But, as expected, the spectra change with the variation of the pH value. This is shown in Fig. 20. It can be seen that a right handed α-helix structure, which is characterized by a maximum at 191 nm and two minima at 210 and 222 nm, is present at pH 9. A sketch of the polymer conformation is given in Fig. 17. The circular dichroism spectra change gradually when lowering the pH and adopt the form of a random coil at pH 3.7, the spectrum of these is characterized by a minimum at about 201 nm and a maximum at about 220 nm. The crossover wavelength remained at 204 nm. It should be noted that this crossover wavelength agreed well with that of the pH-induced helix-coil transition for poly(L-lysine) and PEO-PLL reported in the literature [145,

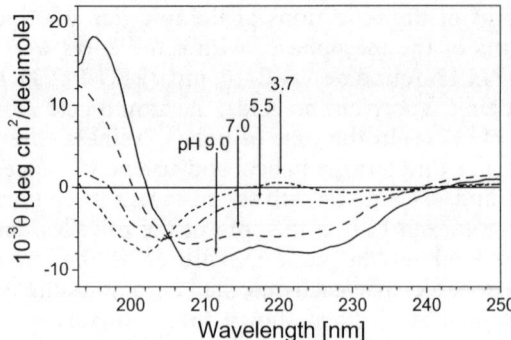

Fig. 20 Circular dichroism spectra of PEO-PLL18 retinoate as a function of the pH value. Reprinted with permission from [153]. Copyright 2000 American Chemical Society

174, 175]. Contributions of β-sheet structures to the spectra were not found to be present. It was concluded that the PLL conformation in the PEO-PLL18 retinoate and PEO-PLL30 retinoate micelles adopts an α-helix for pH values higher than 9.0 and a random coil at a pH lower than 3.7, while between these limiting values a mixture of α-helical and random coil structures is present.

According to Greenfield and Fasman [175], the content of α-helices was estimated from the mean molar ellipticity at a wavelength of 222 nm. It can be seen in Fig. 21 that the α-helix content decreases gradually in the range from pH 9 (95%) to pH 7 (70%). This is followed by a steep decrease occurring between pH 7 and pH 5, from 70 to 20% and approaches zero at about pH 3.7. The stability of the α-helix within the complexes is remarkable when bearing in mind that the helix-coil transition of poly(L-lysine) homopolymers is found at pH 10.3 and 25 °C where the polymer is 35% charged [176].

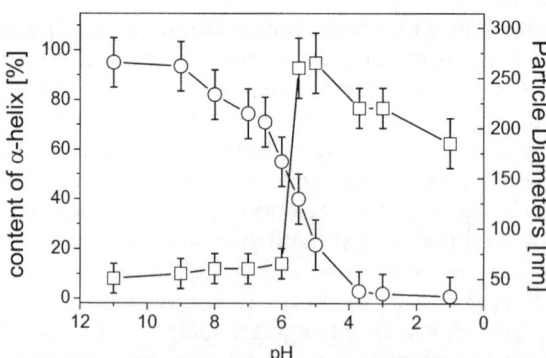

Fig. 21 Variation of the content of the complex in an α-helical state (*circles*) and the particle diameters of the PEO-PLL18 retinoate (*squares*) depend on the pH value. Reprinted with permission from [153]. Copyright 2000 American Chemical Society

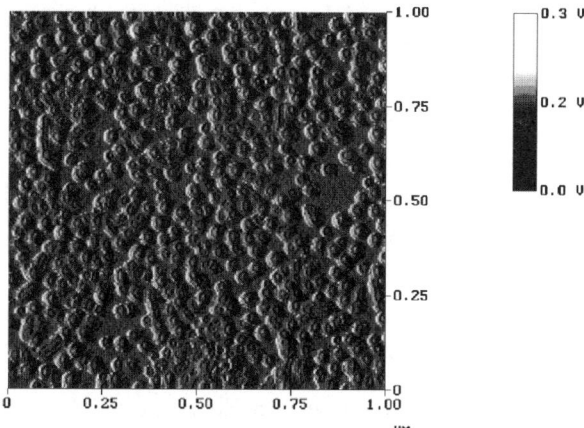

Fig. 22 AFM amplitude picture of a PEO-PLL30 retinoate dispersion (pH 9) dried on mica. Reprinted with permission from [153]. Copyright 2000 American Chemical Society

It was assumed that the α-helix of the complexes is stabilized, firstly by the PEO corona, and secondly by protecting retinoate molecules. This process ends when the retinoate moieties become protonated. This assumption is supported by the pK_a of retinoic acid which is in the range of 6 to 8 and depends strongly on its surroundings, concentration and microenvironment [177, 178]. In order to characterize the size of the micelles, samples of micellar solutions were stored and analyzed by dynamic light scattering directly after the circular dichroism measurement. It was found that the mean particle size was dependent on the pH value. The particle sizes of PEO-PLL18 retinoate and PEO-PLL30 retinoate are essentially the same. At a high pH, where the α-helical conformation of the PLL is predominant and where the retinoic acid has a deprotonated form, the smallest particle sizes are observed. In the range form pH 11 to pH 6 the mean particle diameter increases slightly from 50 nm to 60 nm. Then at pH 5.5 the mean diameter increases to values higher than 250 nm and it decreases constantly down to values of about 180 nm at a pH of 1.

The pH dependency of the particle sizes was confirmed by AFM measurements. An example of PEO-PLL30 retinoate at pH 9 is shown in Fig. 22. The core-shell morphology leads to significantly different AFM amplitudes at the center of the micelles than those in their corona. Exact measurements of the core diameters are not possible on the basis of the AFM data due to the softness of the PEO but the micellar sizes are relatively uniform and agree essentially with the correlation lengths determined by small-angle X-ray scattering. It is interesting that the particle sizes increase considerably within a small range of variation of the pH, namely from 6.0 to 5.5. This is exactly in the region where the content of α-helical conformation decreases most rapidly. The interpretation is that the retinoic acid is bound ionically to the

PLL segments when the pH is higher than 6. This leads to small core-shell micelles with a compact core. At a pH of 5.5 the ionic bonds are broken, the micellar core dissolves or the distinct core-shell structure becomes more diffuse in the sense that a broad core-shell transition develops. Due to the low *cmc* of retinoic acid of about 2×10^{-6} mol/L (pH 7) in aqueous environment, it is probable that the retinoic acid clusters within the micelles and does not diffuse into the aqueous surroundings. The micelle is the most lipophilic environment for the retinoic acid in its protonated form. On the other hand, the dissolution of the compact micellar core into a number of hydrophobic fragments could explain the considerable increase in the particle sizes. It was found that the largest particle sizes are found around pH 5. When the pH is lowered the particle sizes decrease significantly. This is probably due to a rearrangement of retinoic acid clusters due to the lowering of the pH.

In conclusion, it was found that complexes of poly(ethylene oxide)-*b*-poly(L-lysine) with retinoic acid with short poly(L-lysine) segments of 18 and 30 monomers form core shell micelles. The cores of the micelles contain a lamellar smectic A-like structure, formed by a poly(L-lysine) retinoate complex, which is surrounded by a corona of poly(ethylene oxide). Although the poly(L-lysine) chains are relatively short, they adopt an α-helical conformation to a pH as low as 9. This effective stabilization of the α-helix structure seems to be due to the formation of a protective surrounding coat of retinoate and a shell of poly(ethylene oxide).

3.3.3
Polyethyleneimine [179]

The immobilization of retinoic acid by branched PEI with different molecular weights (6 10^2 to 2×10^6 g/mol) is described in the following section. The main interest is on the formation of mesomorphous structures, the release of retinoic acid from its complexes and the formation of nanoparticles.

3.3.3.1
Nanostructures

The optical micrographs of the PEI-retinoate complexes observed between crossed polarizers show that all complexes are highly birefringent, with textures that are independent of their molecular weights [179]. Similar textures are known for some lyotropic lamellar systems of phospholipids [180]. They originate from bilayers that were initially horizontal and corrugate in one or two dimensions to form domes and basins, and each appears as a 'Maltese cross'. In contrast, typical textures of elongated crystals are found in the micrograph of non-complexed all-*trans*-retinoic acid when it is prepared in the same way as the complexes. It was found that a ratio of amino functions to carboxylic acid groups of about 2:1, or higher, is necessary to prevent the crystallization of parts of the retinoic acid. This finding is in agreement with earlier studies in which it was reported that only the primary and the sec-

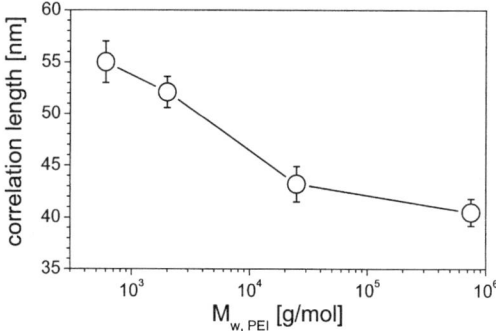

Fig. 23 The correlation length of the lamellar mesophase of the PEI-retinoate complexes depending on the molecular weight of the PEI. Reprinted with permission from [179]. Copyright 2000 American Chemical Society

ondary amino functions could be complexed by surfactants but not the tertiary groups [181]. The amount of the tertiary groups is around 25% in the PEIs used for complexation. This explains why an excess of amino functions, with respect to the carboxylic acid groups, has to be achieved for a complete complexation of the retinoic acid and to avoid crystallinity. Small-angle X-ray scattering measurements, which were carried out on films with thicknesses in the range of 0.05 mm to 1 mm, were used for a quantitative comparison of the complexes. An intense Bragg reflection was found in the scattering curves of all complexes at a scattering vector of about 0.3 nm^{-1}. The shapes of the curves were independent of the thicknesses of the films but they vary with the molecular weight of the PEI. Although no higher order reflections were observed for the complexes in the bulk material, together with the results from polarization microscopy and the fact that in earlier studies only lamellar structured complexes of retinoic acid were found [140] a lamellar structured mesophase for the PEI-retinoate complexes was proposed This assumption will be confirmed later in the section on thin films. The lamellar spacings increase slightly with the increasing molecular weight of the PEI and were determined to be 3.3 nm (PEI-600), 3.4 nm (PEI-2000), 3.4 nm (PEI-25000), and 3.5 nm (PEI-750000). This indicates higher ordered structures of the complexes with the lower molecular weight PEIs than the complexes with higher molecular weight PEIs.

The correlation lengths of the complexes, which were determined from the widths of the reflections, decrease with increasing molecular weight, from about 55 nm to 41 nm (see Fig. 23). This is a further indication for a higher order of the complexes with the PEI of low molecular weight. A probable reason for this influence of the molecular weight PEI on the supramolecular order is that a higher molecular weight would lead to higher packing constraints than a lower molecular weight would do. In addition to the reduction of the order of the lamellar structures this may lead to different degrees of frustrations [15]. The relations between the order and release profiles of retinoic acid from the complex films will be discussed later. In

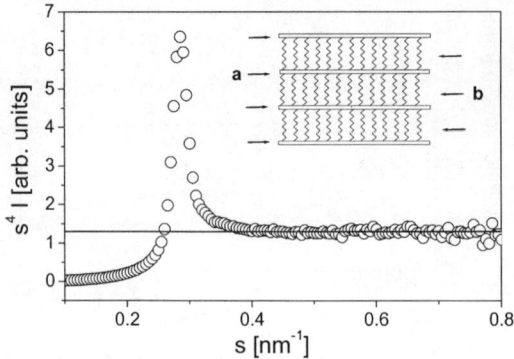

Fig. 24 Small-angle X-ray scattering intensity in a $s^4I(s)$-s-plot (*circles*). The *solid line* represents the best fit according to Porod's law at scattering vectors in the range of 0.4 to 0.8 nm^{-1}. The *insert* shows an idealized model of smectic A-like PEI retinoate structures, consisting of well-separated supramolecular lamellar sheets: (*a*) ionic layers with thicknesses of 0.6 nm (PEI-600), 0.7 nm (PEI-2000 and PEI-25000) and 0.8 nm (PEI 750000); (*b*) nonionic layers with thicknesses of 2.7 nm (PEI-600, PEI-2000, PEI-25000 and PEI-750000). Reprinted with permission from [179]. Copyright 2000 American Chemical Society

earlier works on polyelectrolyte surfactant complexes it was found that a good model for the description of their mesophase structures could be made by placing the microphase separation into ionic and hydrophobic regions. Often the density transition between these regions is sharp and is similar to that found for strongly segregated block copolymers. Sharp phase boundaries are identified by the application of Porod's law [182, 183]. The values of $s^4I(s)$, as shown in Fig. 24, were found to be constant for a scattering vector in the range of 0.4 to 0.8 nm^{-1}. This proves that the structures of the complexes are consistent with Porod's law. A broader transition or a statistical structuring of the domain boundary, as typically observed in microphase-separated block copolymers [182], can be excluded. Small deviations from a sharp boundary would indicate a significant deviation from Porod's law [183]. Therefore it was concluded that the phase boundaries of the complexes are of the order of 1 to 2 atomic distances. Using Eq. (1.2) in Ref [1], the average chord lengths were calculated to be 0.93 nm (PEI-600), 1.06 nm (PEI-2000), 1.10 nm (PEI-25000), and 1.22 nm (PEI-750000). These values are similar to the polymeric complexes of siloxane surfactants investigated earlier [184]. The simplest complex structure, which is in agreement with these numbers is a microphase-separated model constructed from a smaller ionic phase (polyelectrolyte plus ionic head groups) and a larger nonionic phase (hydrophobic moieties). The slight increase of l_p with the increasing molecular weight of the PEI is consistent with the assumption that the packing of the structure of the complexes is better for the low molecular weight complexes than for the high molecular weight complexes. The thicknesses d_1 and d_2 of the lamellae within the smectic A-like structures of the complexes were calculated using Eq. (1.12) in Ref [1]. The values are d_1=0.6

nm, d_2=2.7 nm (PEI-600), d_1=0.7 nm, d_2=2.7 nm (PEI-2000), d_1=0.7 nm, d_2=2.7 nm (PEI-25000) and d_1=0.8 nm, d_2=2.7 nm (PEI-750000). d_1 was assigned to the ionic lamellae, which are enriched in the polyelectrolytes plus the carboxylic head groups, and d_2 to the nonionic lamellae, which are enriched in the retinoic acid tails. As expected, the value of d_1 increases slightly with the molecular weight of the PEI, while the value of d_2 do not depend on the weight of the PEI used for complexation. A sketch of the structure is given in Fig. 24 (insert).

3.3.3.2
Thin Films

Thin films of the complexes on silicon wafers were prepared by the spin-coat technique from solutions of the complexes. The film structures were investigated by X-ray reflectivity. Reflectivity curves of thin films are shown in Fig. 25. Similarly to complexes of a fluorinated surfactant [1], well-defined double-layer stacks developed within a few seconds simply as a result of the deposition of droplets of the solutions of the complex. The molecular weight of the PEI increases in the line from curve a to d. The presence of Kiessig fringes in curves a and b indicate smooth films for the complexes PEI-600-retinoate and PEI-2000-retinoate. No Kiessig fringes were observed for the complexes with the PEI of the higher molecular weights (curves c and d). A clear second order maximum is present in the reflectivity curve of PEI-600-retinoate, which proves the assumptions of a lamellar structure of the complex. The long periods of the structures in the films are 3.3 nm (PEI-600), 3.4 nm (PEI-2000), 3.4 nm (PEI-25000) and 3.6 nm (PEI-750000). Within the experimental errors these values are in good agreement with those determined for the structures of the complexes in bulk form.

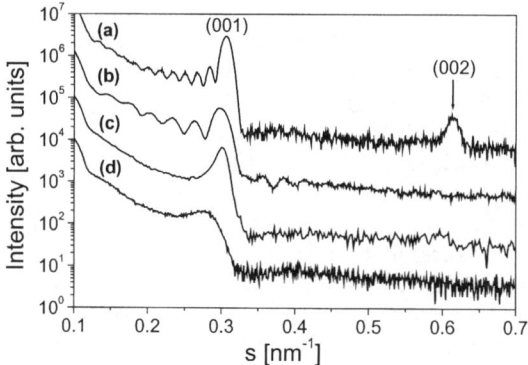

Fig. 25 X-ray reflectivity curves of thin films of PEI-retinoate complexes at the silicon wafer surfaces. The molecular weights of the PEI used for complexation are 600 g/mol (a), 2000 g/mol (b), 25000 g/mol (c) and 750000 g/mol (d). Reprinted with permission from [179]. Copyright 2000 American Chemical Society

Because the same mesomorphous structure was developed in a slow film forming process (solvent casting) and in a fast film forming process (spin coating), it was concluded that the time which is necessary for the formation of the mesophase is shorter than the time necessary for the formation of films. The latter is in the order of one second. The thicknesses of the films formed by the lower molecular weight PEIs were 80±2 nm (PEI-600) and 47±2 nm (PEI-2000) as calculated from the angular positions of the fringes [185]. The reflectivity curves were explained as resulting from about 25 (PEI-600) and 15 double layers (PEI-2000) respectively. Here the double layer is defined as a structure building block with an ionic sheet covered by nonionic sheets. The latter are formed by the tails of the retinoic acid. In contrast to the complexes of the lower molecular weight PEI, the film surfaces obtained by the complexes of the higher molecular weight PEIs are not sufficiently smooth to produce Kiessig fringes in their reflectivity curves. The correlation length of the lamellar structure in the direction vertical to the wafer surface was determined, from the width of the Bragg peaks, to be 66±5 nm (PEI-25000) and 32±5 nm (PEI-750000). The structure of the films can be explained as consisting of multi-layers aligned parallel to the wafer surface.

The observation of the macroscopic orientation of the complexes to multi-layers with a variability in their order which were produced in a single-step procedure represents a significant step in the progress of using self-assembly for the preparation of thin organic films. Mechanical fields, present at the spin coating procedure, are probably the reason for the macroscopic orientation of the lamellar stacks parallel to the underlying substrate surface. Such films, when highly loaded with an effective drug, may be useful as a colloidal pharmaceutical formulation.

3.3.3.3
Release Properties

Films of the complexes are stable in water at a pH of 7 while they dissolve at pH 5. This can be explained by the pK_a value of retinoic acid, which is, for example, 6.05 in 150 mM NaCl and 6.49 in 5 mM NaCl [163]. Therefore, the anionic retinoic moieties within the complexes will be protonated at pH values lower than the pK_a which lead to the cleavage of the ionic bonds in the complexes. The first experiments to evaluate the release properties of retinoic acid from thin films of the complexes were performed by using FTIR and surface tension measurements. Films were immersed in solutions of 0.15 m sodium chloride at pH 5 for both methods. The increase of the absorbance at 1255 cm^{-1} (C-O stretch vibration) [186] in the FTIR spectra was used as a qualitative measure for the release of retinoic acid from the PEI-retinoate complexes. For comparison, the spectra of the complex and of the non-complexed retinoic acid are shown at wave numbers around 1255 cm^{-1} (Fig. 26, insert curves a and b). The time-dependency of the absorbance, which is a relative measure of the amount of released retinoic acid, is shown in Fig. 26. It can be seen that the increase of the absorbance, and therefore the release

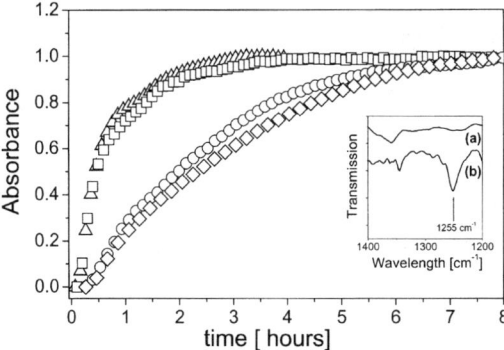

Fig. 26 Time-dependency of the FTIR spectra of the PEI retinoate complexes at a wave number of 1255 cm^{-1}. The samples are PEI-600 retinoate (*diamonds*), PEI-2000 retinoate (*circles*), PEI-25000 retinoate (*squares*) and PEI-750000 (*triangles*). The *insert* shows the transmission of a complex (*a*) and of non-complexed retinoic acid (*b*) around a wavelength of 1255 cm^{-1}. Reprinted with permission from [179]. Copyright 2000 American Chemical Society

of retinoic acid from the complexes, increases in the line PEI-600, PEI-2000, PEI-25000, PEI-750000.

Obviously the release from the complexes with the higher molecular weight PEIs is faster than that from the complexes with the lower molecular weight PEIs. Because the ratio of primary to secondary to tertiary amino functions in all PEIs is approximately the same, it was suggested that the different release profiles probably originate from the different supramolecular ordering of the complexes and are not due to the minor differences in their molecular composition. Together with the results from the X-ray reflectivity measurements it has been concluded that the release of retinoic acid from a complex film is relatively slow for the higher ordered films and relatively fast for the less ordered films. This explains why the release from the complexes with higher molecular weight PEIs is faster than it is from the complexes with the lower molecular weight PEIs.

The molecular structure of retinoic acid is typical for an amphiphilic compound that is concentrated at interfaces. Further, the carboxylic acid groups allow such compounds to adjust their amphiphilic character by the degree of their dissociation. Surface tension measurements were carried out in order to determine the surface activity of retinoic acid [179]. The surface tension with respect to the concentration at pH 5 decreases more strongly than at pH 9. This reflects the fact that the protonated form of retinoic acid is more efficient in its surface activity than the deprotonated form. The critical micelle concentrations are 3.7±0.5 mg/L (pH 5) and 19±2 mg/L (pH 9). The limiting surface tension values in both curves is about 35 mN/m. Due to the precipitation of retinoic acid, the highest concentration in the surface tension curve at a pH of 5 was 20 mg/L. By contrast the solubility at pH 9 is at least 1 g/L. In order to verify the results from the FTIR measurements, films of the complexes were immersed in a solution of 0.15 mol/L sodium

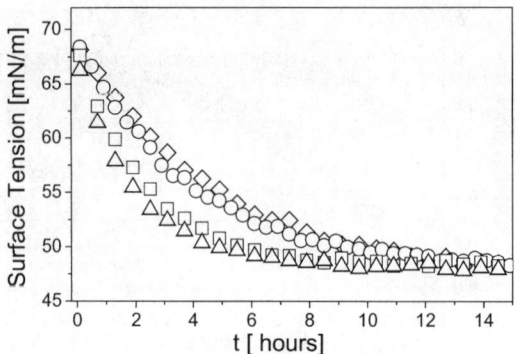

Fig. 27 The effect of the molecular weight of PEI retinoate complexes on the surface tension of 0.15 m/L sodium chloride solutions as a function of time. The samples are PEI-600 retinoate (*diamonds*), PEI-2000 retinoate (*circles*), PEI-25000 retinoate (*squares*) and PEI-750000 retinoate (*triangles*). Reprinted with permission from [179]. Copyright 2000 American Chemical Society

chloride at pH 5. It can be seen in Fig. 27 that the surface tension decreases with time when the films were inserted in the solutions. The decrease is steeper the higher the molecular weight of the PEI is. The final value of the surface tension for all films is around 48 mN/m. This corresponds to a maximum concentration of about 2 mg/L, which is below the critical micelle concentration for retinoic acid. We interpret the reduction of the surface tension to be due to the release of retinoic acid from the films. The release is faster for the higher molecular weight PEI complexes than for the lower. Because the limiting values of the surface tensions are independent of the molecular weight of the PEIs, it was concluded that the kinetics of the release are affected by the PEI used, but not the overall amount of released retinoic acid. This is consistent with the results from the FTIR spectroscopy, and with the assumption that the complexes differ only in their supramolecular orders.

3.3.3.4
Nanoparticles

The formation of the nanoparticles of the complexes was carried out with the aid of poloxamer 188, a triblock copolymer, consisting of two polyethyleneoxide blocks and a polypropyleneoxide block (see Fig. 12). The diameters of the particles, as determined by dynamic light scattering, are shown in Fig. 28. Surprisingly, the diameters decrease with the increasing molecular weight of the PEI as follows: 580 nm (PEI-600), 240 nm (PEI-2000), 185 nm (PEI-25000), 160 nm (PEI-750000). The polydispersities are about 0.25. The sizes were checked by electron microscopy (not shown). Park and Choi [187] have shown that the structures of the PEIs are randomly branched but they become more compact with increasing molecular weight. Probably the

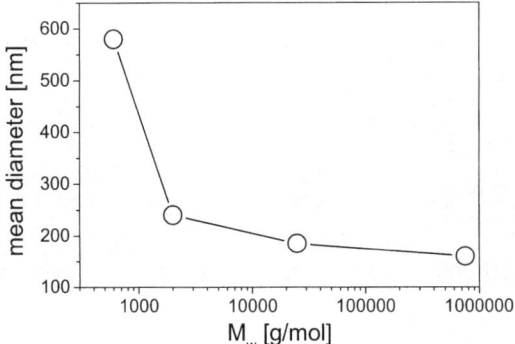

Fig. 28 The molecular weight dependency of the mean diameter of particles of the PEI-retinoate complexes. The molecular weight is the weight average of the PEI used for complexation. Reprinted with permission from [179]. Copyright 2000 American Chemical Society

unexpected size-dependency of the nanoparticles reflects the structural properties of branched PEI. On the basis of this hypothesis it must be expected that PEI-750000 retinoate forms a significantly more compact particle structure than PEI-600 retinoate. AFM measurements, carried out on dried PEI retinoate particles, confirmed the expectation of an increase in compactness of the particles with increasing molecular weight. Examples are shown in Fig. 29. It was found that the higher the molecular weight of the PEI, the more compact the particles. The height profiles of PEI-750000 retinoate (Fig. 30b), for example, are those of typical rigid spheres. The diameters of the particles determined by AFM are about 50 nm smaller than those determined by dynamic light scattering. The higher radii in solution can be explained as an effect of the Poloxamer corona that surrounds the particles, which is swollen in aqueous solution, leading to higher radii in solution than in the dry state. The AFM pictures of the PEI-600 retinoate nanoparticles are unique in their appearance. Here, the particles are collapsed and surrounded by crystalline Poloxamer 188. For comparison the pure Poloxamer, which was prepared from aqueous solution on mica in the same way as the particles, was investigated. It can be seen in Fig. 29d that Poloxamer 188 forms structures which are very similar to those found in the corona of the particles in Fig. 29a. A characteristic detail of the PEI-600 retinoate particles is that there are holes in the centers of many of them. These are indicated in the 2×2 µm large bird's eye view of Fig. 29b, pointed by arrows. A typical cross-section profile is shown in Fig. 29a. Obviously, the PEI-600 retinoate particles in solution were not homogeneous spheres. Probably these particles may have a doughnut-shape-or toroid-structure in solution, such as described recently by Förster et al. for polyelectrolyte block copolymer micelles [188]. A different morphology of the PEI-600 retinoate particles can explain why the mean diameter of the particles is significantly higher than that of the other complexes.

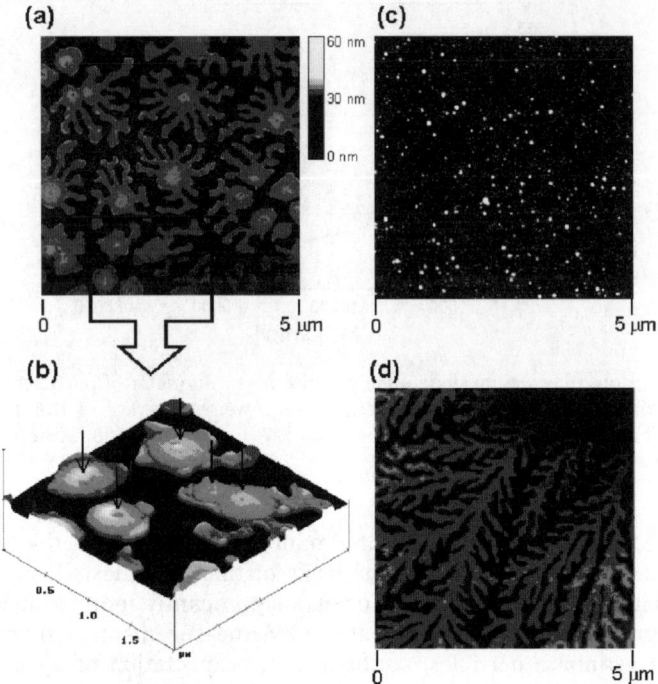

Fig. 29 AFM images of PEI-retinoate particles dried on mica surfaces. (**a**) and (**c**) are top-view images of the PEI-600 retinoate and the PEI-750000 retinoate, respectively. The bird's-eye view (**b**) is a magnification of (**a**) enclosed in the *plotted rectangle*. *Arrows* in (**b**) indicate the depressions in the center of the particles. (**d**) is the top view of Poloxamer 188 after the aqueous solution had been dried out. Reprinted with permission from [179]. Copyright 2000 American Chemical Society

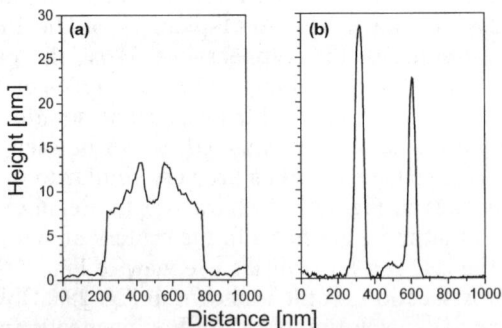

Fig. 30 Height profiles of the cross section of nanoparticles formed by PEI-600 retinoate (**a**) and PEI-750000 retinoate (**b**). Reprinted with permission from [179]. Copyright 2000 American Chemical Society

In conclusion, it was shown that the complexes formed by PEI and the amphiphilic drug retinoic acid are lamellar structured materials, which can be processed as lamellar layered ultra-thin films as well as nanoparticles. The film forming properties of the complexes are the better the lower the molecular weight of the PEI is. All complexes are stable in physiological saline at pH 7, whereas retinoic acid is released at pH 5. The higher the molecular weight, the faster its release. With the help of a water-soluble block copolymer, nanoparticles were formed, the size of which decreases with the increasing molecular weight of the PEI. This was explained as a phenomenon induced by the pristine PEI structure. The shapes of the particles from the complexes of the high molecular weight PEI are compact spheres, whereas those of the lowest molecular weights are probably of doughnut-shape or toroid structure.

4
Theory of Polyelectrolyte Complexation

Although the experimental body of work on polyelectrolyte complexation is very large, there are very few quantitative theoretical approaches. If the polyelectrolytes are weakly charged which is not the case in most of the experiments discussed in this review, the electrostatic interactions are weak and a theory similar to the Debye-Hückel theory of electrolytes can be proposed. This was first done by Borue and Erukhimovich [189]. In this short section, we review the results that have been obtained for the phase diagram of symmetric polyelectrolyte complexes [190], for polyelectrolyte multilayers [191] and for block-polyampholytes which are diblock copolymers with one sequence carrying positive charges and one sequence carrying negative charges [192]. Finally we briefly describe the concept of the effective interaction between two polyelectrolyte complexes.

4.1
Debye-Hückel Theory of Polyelectrolyte Complexes

In a simple mean field theory, the charge density in an electrolyte solution vanishes and the electrostatic contribution to the free energy of the solution also vanishes. The electrostatic free energy is therefore an effect of the charge density fluctuations. The starting point of the Debye-Hückel theory of electrolytes is to assume that these fluctuations are small and can be treated as an expansion which is limited to quadratic order in the free energy. We use here the same description for polyelectrolyte complexes. It is consistent only if the fractions of charged monomers f_+ and f_- on the two polymers forming the complex are small. A polyelectrolyte complex in solution is at least a 5 component system comprising the solvent, the two polymers at concentrations c_+ and c_- and their respective counterions at concentrations n_- and n_+. We consider as counterions any small ions in the solution; if there is added salt, its ions have the same chemical nature as the polymer counterions. The only constraint on the concentrations is then the macroscopic electroneutrality of the solution, $n_+ + c_+ = n_- + c_-$. For a sake of simplicity, we consider here only symmetrical complexes where the same very large number of monomers N, the same fraction of charged monomers f and the same concentrations, $c_+ = c_- = c$ and $n_+ = n_- = n$.

The calculation of the electrostatic free energy is done by expanding the solution free energy in a given configuration at lowest order in powers of the 4 concentration fluctuations and then by summing over all possible concentration fluctuations. As an example, we discuss first here the simple case where the added salt density is large (n>>c). We define the Debye Hückel screening length counting only the small ions as $\kappa^2 = 8\pi \, l_B n$ where $l_B \, q^2/(4\pi\varepsilon kT)$ is the so-called Bjerrum length. The electrostatic interactions are attractive and provoke the precipitation of the complex. The precipitation gives a phase separation between the complex phase rich in polymer and a dilute phase containing essentially no polymers but containing small ions (salt). The results are very similar to the standard Debye-Hückel theory, the electrostatic osmotic pressure difference between the complex and the dilute phase can be written as $\Pi_{el} = -kT/\xi_{el}^3$ where ξ_{el} is the electrostatic correlation length given by

$$\xi_{el} \sim \frac{\kappa a}{(4\pi l_B f^2 c)^{1/2}} \tag{3}$$

At equilibrium the osmotic pressure is equal in the two phases and the electrostatic osmotic pressure is balanced by the excluded volume osmotic pressure in the complex that for Gaussian polymers (at the θ point) reads $\Pi_{ev} = w^2 c^{3/2}$. The concentration in the complex then reads

$$c_{comp} \sim \frac{f^2}{a^2 n \, w^{3/4}} \tag{4}$$

It increases with the charge of the polymers and as expected, decreases with the ionic strength. So far, we have ignored the translational entropy of the polymer chains that tends to zero in the limit where N tends to infinity. If N is smaller, the translational entropy of the polymer chains must be included in the pressure balance and stabilizes a homogeneous solution when it becomes of the order of the electrostatic attractive pressure. Complex formation thus only occurs if

$$\frac{n^2 a^2 w^{3/4}}{N f^4} < 1. \tag{5}$$

If the molecular weight is too small or if the fraction of charged monomers is too low, the mixture of positively charged and negatively charged polyelectrolytes is soluble in water.

A more detailed study of the phase diagram of the polyelectrolyte complex must take into account the non-electrostatic interactions between the two polymers characterized by the Flory interaction parameter χ. This inter-

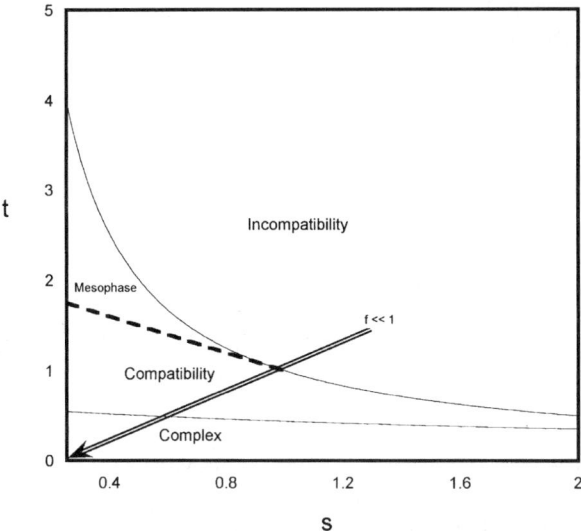

Fig. 31 Theoretical phase diagramm for a symmetric mixture of positive polyelectrolytes. s measures the effect of ionic strength and t that of the incompatibility

action most often leads to a macroscopic phase separation into two phases each containing mostly one of the polymers. There is thus a competition between a complexation transition and a demixing transition. In order to have a qualitative idea of the phase diagram, we have calculated the concentration correlation matrix in the solution at the same level of approximation (the so-called one-loop approximation in the field theory language). The signature of a phase transition is a divergence of this matrix. For the complexation transition, the order parameter is the total polymer concentration $c_+ + c_-$ and for the demixing transition it is the charge density $\rho = c_+ - c_-$. The spinodal line associated to the phase transitions occurs when the corresponding correlation function diverges in the macroscopic limit of zero wave vector. We also looked for divergences at finite wave vectors that corresponds to the formation of mesophases in the solution where the polymer composition of the solution oscillates periodically with space. The theoretical phase diagram for a symmetric mixture of positive polyelectrolyte is presented in Fig. 31. The two dimensionless variable are

$$s = \left(\frac{4\pi l_B}{f^2 c}\right)^{1/2} n\, a, \quad t = \frac{\chi}{a(4\pi l_B f^2 c)^{1/2}}, \tag{6}$$

s measures the effect of ionic strength and t that of the incompatibility, all other quantities being kept constant. There are 4 phases, a complex phase at low values of χ, a demixing region at high values of χ, a homogeneous solution, and a mesophase region at small ionic strength and intermediate χ. If the fraction of charged monomers is increased as in the experiments of Djadoun et al. [193], at low fraction of charged monomers, the solution demixes into two phases it then becomes homogeneous and at large fraction of charged monomers, it forms a complex as indicated by the arrow.

4.2
Polyelectrolyte Multilayers

Over recent years, a major development in the field of polyelectrolyte complexes has been the introduction of polyelectrolyte multilayers by Decher [194]. Some of their properties are described in this paper and they have become one of the very efficient and versatile ways of making surface coatings. We very briefly discuss here the implications of the Debye-Hückel properties of polyelectrolyte complexes for the formation of polyelectrolyte multilayers.

The first layer is obtained by adsorbing a polyelectrolyte layer (say with a positive charge) on an oppositely charged substrate (a negative substrate). One wants to strongly inverse the substrate charge upon adsorption [195] and this may require a non-electrostatic attraction between polyelectrolyte and substrate. The first layer is, for this reason, often made of another polyelectrolyte than the rest of the build up. We assume here that the structure of this adsorbed layer is the thermodynamic equilibrium structure and that it is made at a given high ionic strength (a salt concentration n). For the rest of the build up, we will suppose that the multilayer is never dried and that it is always in contact with an ionic solution at the same ionic strength. This is not how the layers are made in practice but this insures that the polymer remains liquid and close to thermodynamic equilibrium at each step. We also assume that the adsorption is irreversible in the sense that the adsorbed polymer amount remains constant but that the polymer chain conformations at each step can equilibrate. The polyelectrolyte is then in a constrained thermodynamic equilibrium. We call Γ_1 the adsorbed monomer amount in the first layer per unit area.

After rinsing in a salt solution, the first layer is put in contact with a dilute solution of polyelectrolyte of opposite charge at the same ionic strength (negatively charged). The negatively charged polymer in the solution forms a complex that neutralizes the already adsorbed positively charged polymer (as above, we only consider the symmetric case where the two polymers have the same degree of polymerization and the same fraction of charged monomers). The density of this complex is equal to the density c_{comp} of a symmetric polyelectrolyte complex in solution. However there is an excess of negatively charged polyelectrolyte in the solution and the negatively charged polyelectrolyte can form loops dangling from the multilayer into the solution. The loop structure is very similar to that of an adsorbed polymer layer and the monomer concentration in this loop layer decays as $c(z) \sim 1/(z+d)^2$ from the edge of the multilayer. The charge of this loop layer overcompensates the charge of the originally adsorbed polymer, it allows for a new charge inversion and can serve to complex the following layer. The quantity of monomers in this non compensated layer is

$$\Delta \Gamma \sim \frac{1}{w^{2/3}} \left[1 - \frac{f^2}{n^{3/2} l_B^{1/2} a^2 w^{2/3}} \right]. \tag{7}$$

The build up process can the be iterated and at each step the amount of polymer added to the layer is of order $\Delta \Gamma$. It increases weakly with ionic strength. This is in qualitative agreement with the experiments of Laddam et al. [196] that were performer in conditions that are as close as possible as the one described here.

This approach is a purely thermodynamic approach in the sense that locally, the polymer is at thermal equilibrium. The density inside the layer is homogeneous and is equal to the complex density c_{comp}. One of the important experimental results is that, in the multilayers, each layer strongly intermixes with its neighbors but that it keeps its identity in the sense that it does not mix with layers far apart. This "freezing" of the structure must be due to an extremely slow diffusion between the layers that our model is not able to study.

4.3
Block Polyampholytes

Another interesting application of the complexation model is to block polyampholytes. These are diblock copolymers composed of a positive polyelectrolyte block of N_P monomers and a negative polyelectrolyte block of N_M monomers. If the two blocks are symmetric (same molecular weight and same fraction of charged monomers), a copolymer solution precipitates as a standard polyelectrolyte complex, the connection between the two chains playing only a small role. If the copolymer is very asymmetric, $N_P \gg N_M$ it can form aggregates such as spherical micelles. The core of the micelle is made of the negative block and of N_P monomers of the positive block and has a concentration c_{comp}; the corona of the micelles is made of the $N_P - N_M$ remaining monomers, it behaves as a polyelectrolyte corona [197]. The geometrical characteristics of these micelles can then be calculated using a standard theory of micellization that estimates independently the free energies of the two blocks. We find an aggregation number

$$p \sim \frac{N_M^{3/5} f^{16/5}}{[n^2 w^{2/3} a^4]^{4/5}} \tag{8}$$

It increase with the charge fraction f and decreases with ionic strength. This is consistent with experiments on block polyampholytes where the pH is varied [198].

4.4
Effective Interaction Between Two Polyelectrolyte Complexes

Let us finally discuss the concept of effective interactions between two polyelectrolyte complexes which is particularly useful if the architecture of the complexes is known a priori, for a review, see [199, 200]. One typically starts with two complexes choosing their center-of-mass distance r as the statistical variable which is kept fixed while all other statistical degrees of freedom such as solvent molecules, counter- and salt ions and monomers are integrated out. The resulting quantity $V(r)$ is the (pair) potential of mean force between two complexes which can be used as a further input in a simple many body theory for many complexes at finite concentration. There have been several attempts to calculate the effective pair interaction theoretically, the most widely known is the classical Debye-Hückel theory of counterion screening resulting in screened Coulomb interactions between the charged monomers [201]. Following the standard approach of Oosawa [202], these can be combined with Flory arguments for polymer flexibility, polymer self-avoidance, and counterion condensation along the charged monomers. Another important contribution comes from the translational entropy of the microions.

A variational theory which includes all these different contributions was recently proposed and applied for completely stretched polyelectrolyte stars (so-called "porcupines") [203, 204]. As a result, the effective interaction $V(r)$ was very soft, mainly dominated by the entropy of the counterions inside the coronae of the stars supporting on old idea of Pincus [205]. If this pair potential is used as an input in a calculation of a solution of many stars, a freezing transition was found with a variety of different stable crystal lattices including exotic open lattices [206]. The method of effective interactions has the advantage to be generalizable to more complicated complexes which are discussed in this contribution-such as oppositely charged polyelectrolytes and polyelectrolyte-surfactant complexes-but this has still to be worked out in detail.

References

1. Thünemann A (2002) Prog Polym Sci 27:1473
2. Kötz J, Kosmella S, Beitz T, Prog Polym Sci (2001) 26:1199
3. Bertrand P, Jonas A, Laschewsky A, Legras R (2000) Macrom Rapid Comm 21:319
4. Ober CK, Wegner G. (1997) Advanced Materials 9:17
5. Antonietti M, Burger C, Thünemann A (1997) TRIP 5:262
6. Antonietti M, Thünemann A. (1996) Current Opinion in Colloid & Interface Sci 1:667
7. MacKnight WJ, Ponomarenko EA, Tirrell DA (1998) Accounts Chem Res 31:781
8. Schmidt J, Decher G, Hong JD (1992) Thin Solid Films 210/211:831
9. Decher G (1997) Science 277:1232-1237
10. Lvov YM, Sukhorukov GB (1997) Biologicheskie Membrany 14:229-250
11. Ariga K, Lvov Y, Kunitake T (1997) J Am Chem Soc 119:2224
12. Struth B, Eckle M, Decher G, Oeser R, Simon P, Schubert DW, Schmitt J (2001) European Physical Journal A 6:351

13. Radtchenko IL, Sukhorukov GB, Mohwald H (2002) Colloids & Surfaces A-Physicochemical & Engineering Aspects. 202:127
14. Caruso F (2001) Adv Mat 13:11
15. Antonietti M, Conrad J, Thünemann A (1994) Macromolecules 27:6007
16. Goddard E. D, Colloids Surf (1986) 19:301
17. Hayakawa K., Kwak JCT (1982) J Phys Chem 86:3866
18. Hayakawa K, Kwak JCT (1983) J Phys Chem 87:506
19. Philipp B, Dawydoff W, Linow KJ (1982) Zeitschrift für Chemie 22:1
20. Zezin AB, Kabanov (1982) Usp Khim 51:1447
21. Kabanov VA, Zezin AB (1984) Pure appl Chem 56:343
22. Kabanov VA (1994) In: Dubin P, Bock J, Davies RM, Schulz DN, Thies C (eds) Macromolecular Complexes in Chemistry and Biology. Springer Verlag, Berlin Heidelberg New York, p 151
23. Tsuchida E, Osada Y, Sanada K (1972) J Polym Sci A 10:3397
24. Tsuchida E, Osada Y, Ohno H (1980) J Macromol Sci B 17:683
25. Tsuchida E, Abe K (1982) Adv Polym Sci 45:1
26. Bakeev KN, Izumrudov VA, Kabanov VA (1988) Doklad Akad Nauk SSSR 299:1405
27. Zezin AB, Kabanov, VA (1982) Usp Khim 51(9):1447
28. Izumrudov VA, Kharenko OA, Kharenko AV, Gulaeva ZG, Kasaikin VA, Zezin AB, Kabanov VA (1980) Visokomol Soed A 22:692
29. Pergushov DV, Izumrudov VA, Zezin AB, Kabanov VA (1993) Polymer Science Ser A 35:844
30. Bakeev KN, Izumrudov VA, Kuchanov SI, Zezin AB, Kabanov VA (1988) Doklad Akad Nauk SSSR 300:132
31. Izumrudov VA, Bronich TK, Saburova OS, Zezin AB, Kabanov VA (1988) Makomol Chem Rapid Comm 9:7
32. Dautzenberg, H (2001) In: Radeva T (ed) Physical Chemistry of Polyelectrolytes. Surfactant science series, vol. 99, Marcel Dekker, Inc., p 743
33. Micheals AS, Mir I, Schneider NS (1965) J Phys Chem 69:1447
34. Dautzenberg H, Hartmann J, Grunewald S, Brand, F (1995) Ber Bunsen-Gesell, Conference Proceedings of "Polyelectrolytes Potsdam' 95" 100:1024
35. Kriz J, Dautzenberg H (2001) J Phys Chem 105:3846
36. Buchhammer HM, Lunkwitz K, Pergushov DV (1997) Macromol Symp 126:157
37. Petzold G, Nebel A, Buchhammer HM, Lunkwitz K (1998) Colloid Polym Sci 276:125
38. Brand F, Dautzenberg H (1997) Langmuir 13:2905
39. Karibyants N, Dautzenberg H, Cölfen H (1997) Macromolecules 30:7803
40. Harrison CA, Tan JS (1999) J Polym Sci Part B Polym Phys 37:275
41. Tan JS, Harrison CA, Caldwell KD (1998) J Polym Sci Part B Polym Phys 36:537
42. Hallberg RK, Dubin PL (1998) J Phys Chem B 102:8629
43. Dautzenberg H, Rother G (1988) J Polymer Sci Polym Phys Ed Part B Polymer Physics 26:353
44. Dautzenberg H, Rother G (1992) Makromol Chem Macromol Symp 61:94
45. Dautzenberg H (2000) Makromol Chem Macromol Symp 162:1
46. Chu B (1991) Laser Light Scattering, 2nd edn. Academic Press, Boston
47. Provencher SW (1982) Comp Phys Comm 27:213
48. Philipp B, Dautzenberg H, Linow KJ, Kötz J, Dawydoff W (1989) Progress in Polymer Science 14:91
49. Dautzenberg H, Koetz J, Linow K J, Philipp B, Rother G (1994) In: Dubin P, Bock J, Davies RM, Schulz DN, Thies C (eds) Macromolecular Complexes in Chemistry and Biology. Springer, Berlin Heidelberg New York, p 119
50. Dautzenberg H, Rother G, Hartmann J (1994) In: Schmitz KS (ed) Macro-ion Characterization from Dilule Solutions to Complex Fluids. ACS Symposium Series 548 ACS, Washington DC p 219
51. Papisov IM, Litmanovich AA (1989) Adv Polym Sci 90:139
52. Dautzenberg H, Linow K J, Rother G (1990) Acta Polymerica 41:98

53. Karibyants N, Dautzenberg H (1998) Langmuir 14:4427
54. Dautzenberg H, Rother G, Linow KJ, Phlipp B (1988) Acta Polymerica 39:157
55. Dautzenberg He, Dautzenberg Ho (1985) Acta Polymerica 36:102
56. Dautzenberg H (1997) Macromolecules 30:7810
57. Dautzenberg H, Karibyants N (1999) In: Noda I, Kokufuta E (eds) Proceedings of the 50th Yamada Conference and Second International Symposium on Polyelectrolytes, Inuyama, Japan. Yamada Science Foundation, Osaka, p 284
58. Dautzenberg H, Karibyants N (1999) MacromolChem Phys 200:118
59. Kabanov A V, Bronich TK, Kabanov VA, Kui Yu, Eisenberg A (1996) Macromolecules 29:6797
60. Harada A, Kataoka K (1997) J Macromol Sci-Pure Appl Chem A34(10):2119
61. Dautzenberg H (2000) Macromol Chem Phys 201:1765
62. Zintchenko A, Dautzenberg H, Tauer K, Khrenov V: (2002) Langmuir 18:1386
63. Dautzenberg H unpublished work
64. Schild HG (1992) Prog Polym Sci 17:163
65. Shibayama M, Tanaka T (1992) Adv Polym Sci 109:1
66. Hirose Y, Amiya T, Hirokawa Y, Tanaka T (1987) Macromolecules 20:1342
67. Matsuo ES, Tanaka T (1988) J Chem Phys 89:1696
68. Beltran S, Baker JP, Hooper HH, Blanch HW, Prausnitz JM (1991) Macromolecules 24:549
69. Lee F, Hsu CH (1998) J Appl Polym Sci 69:1793
70. Dautzenberg H, Gao Y, Hahn M (2000) Langmuir 16:9070
71. Xia J, Dubin PL (1994) In: Dubin P, Bock J, Davies RM, Schulz DN, Thies C (eds) Macromolecular Complexes in Chemistry and Biology. Springer, Berlin Heidelberg New York, p 247
72. Kokufuta E (1994) In: Dubin P, Bock J, Davies RM, Schulz DN, Thies C (eds) Macromolecular Complexes in Chemistry and Biology. Springer Verlag, Berlin Heidelberg New York, p 301
73. Tribet Ch (2001) In: Radeva T (ed) Physical Chemistry of Polyelectrolytes. Surfactant science series, vol. 99, Marcel Dekker, Inc., p 687
74. Kabanov VA, Zezin AB, Izumrudov VA (1987) In: Ovchinnikov Yu A (ed) Progress in Science and Technique. Biotechnology. VINITI, Moscow, 4:159 (in Russian)
75. Dautzenberg H, Karibyants N, Zaitsev Syu (1997) Macromol Rapid Commun 18:175
76. Gorokhova I, Zinchenko A, Dautzenberg H, Zaitsev SYu (2002) Langmuir, submitted
77. Czichocki G, Dautzenberg H, Capan E, Vorlop KD (2001) Biotechnol Lett 23:1203
78. Kabanov AV, Felgner PL, Seymour LW (1998) Self-assembling Complexes for Gene Delivery. Wiley
79. Wolfert MA, Dash PR, Nazarova O, Oupicky D, Seymour LW (1999) Bioconjugate Chem 10:993
80. Garnett MC (1999) Therapeutic Drug Carrier Systems 16:147
81. Kabanov AV, Vinogradov SV, Suzdaltseva YG, Alakhov VY (1995) Bioconjugate Chem 6:639
82. Vinogradov SV, Bronich TK, Kabanov AV (1998) Bioconjugate Chem 9:805
83. Read ML, Dash PR. Clark A, Howard K, Oupický D, Toncheva V, Alpar HO, Schacht EH, Ulbrich K, Seymour LW (2000) Eur J Pharm Sci 10:169
84. Kabanov AV (1999) Pharm Sci Technol Today 2:365
85. Oupický D, Koňá, Č, Ulbrich K, Wolfert MA, Seymour LW (2000) J Controlled Release 65:149
86. Bronich T, Kabanov VA, Marky LA (2001) J Phys Chem B 105:6042
87. Dautzenberg H, Zintchenko A, Konak C, Reschel T, Subr V, Ulrich,K (2001) Langmuir 17:3096
88. Shinoda K, Hayashi T, Yoshida T, Sakai K, Nakajima A (1976) Polymer J 8:202
89. Philipp B, Dautzenberg H, Linow KL, Koetz J (1989) Prog Polymer Sci 14:91
90. Koetz J, Kosmella S (1994) J Colloid Interface Sci 168:505

91. Petzold G, Schwarz S, Buchhammer HM, Lunkwitz K (1997) Angew Makromol Chem 253:1
92. Oertel U, Buchhammer HM, Müller M, Nagel J, Braun HG, Eichhorn KJ, Sahre K (1999) Macromolecular Symposia145:39
93. M. Müller (2002) ATR-FTIR Spectroscopy at Polyelectrolyte Multilayer Systems. In: Tripathy SK, Kumar J, Nalwa HS (eds) Handbook of Polyelectrolytes and Their Applications. American Scientific (ASP), pp 293-312
94. Reihs T, Müller M, Lunkwitz K (2002) Colloids and Surfaces (in press, 2002)
95. Reihs T and Müller M (2003, in preparation)
96. Kramer G, Buchhammer HM, Lunkwitz K (1997) Colloids & Surfaces A 122:1
97. Buchhammer HM, Petzold G, Lunkwitz K (1999) Langmuir 15:4306
98. Ladam G, Schaad P, Voegel JC, Schaaf P, Decher G, Cuisinier F (2000) Langmuir 16:1249
99. Borue VY, Erukhimovich IY (1990) Macromolecules 23:3625
100. Castelnovo M, Joanny JF (2000) Langmuir 16:7524
101. Müller M (in preparation 2003)
102. Senger B, Voegel JC, Schaaf P (2000) Coll Surf A 165:255
103. Vuillaume PY, Jonas AM, Laschewsky A (2002) Macromolecules 35:5004
104. Müller M (2001) Biomacromolecules 2:262
105. Müller M, Reihs T, Lunkwitz K (2002) Tagungsband des Makromolekularen Kolloquiums, Freiburg, 21-23 February 2002, pp 58
106. Engelking J, Ulbrich D, Menzel H (2000) Macromolecules 33:9026
107. Pinto MR, Schanze KS (2002) Synthesis-Stuttgart (9 Special Issue SI):1293
108. Müller M, Rieser T, Dubin PL, Lunkwitz K (2001) Macromol Rapid Comm 22:390
109. Müller M, Heinen S, Oertel U, Lunkwitz K (2001) Macromol Symposium 164:197
110. Schwarz B, Schönhoff M (2002) Langmuir 18:2964
111. Müller M, Rieser T, Köthe M, Brissova M, Lunkwitz K (1999) Macromol Symp 145:149-159
112. Goddard ED (1986) Colloids and Surfaces 19:301
113. Zhou SQ, Chu B (2000) Adv Mater 12:545
114. Thünemann AF, Schnöller U, Nuyken O, Voit B (1999) Macromolecules 32:7414
115. Thünemann AF, Schnöller U, Nuyken O, Voit B (2000) Macromolecules 33:5665
116. Thünemann AF, Ruppelt D, Burger C, Müllen K (2000) J Mater Chem 10:13
117. Thünemann AF, General S (2001) Macromolecules 34:6978
118. Ruokolainen J, Makinen R, Torkkeli M, Makela T, Serimaa R, Tenbrinke G, Ikkala O (1998) Science 280:557
119. Antonietti M, Neese M, Blum G, Kremer F (1996) Langmuir 12:4436
120. Zhou SQ, Burger C, Yeh FJ, Chu B (1998) Macromolecules 31:8157
121. Thünemann AF, Lochhaas KH (1998) Langmuir 14:4898
122. Behnke M, Tieke B (2002) Langmuir 18:3815
123. Thünemann AF (1999) Adv Mater 11:127
124. Thünemann AF, Ruppelt D (2001) Langmuir 17:5098
125. Faul CFJ, Antonietti M (2002) Chem A Eur J 8:2764
126. Guan Y, Antonietti M, Faul CFJ (2002) Langmuir 18:5939
127. General S, Antonietti M (2002) Angew Chem Int Ed 41:2957
128. Thünemann AF, General S (2000) Langmuir 16:9634
129. Thünemann AF, General S (2001) J Controlled Release 75:237
130. Hentze HP, Khrenov V, Tauer K (2002) Colloid and Polymer Science, published online 26.7.2002
131. Thünemann AF, Wendler U, Jaeger W, Schnablegger H (2002) Langmuir 18:4500
132. Thünemann AF, Sander K, Jaeger W, Dimova R (2002) Langmuir 18:5099
133. Packer L (ed) (1990) Methods in Enzymology, vol 190. Academic Press, San Diego, CA
134. Lewin AH, Bos ME, Zusi FC, Nair X, Whiting G, Bouquin P, Tetrault G, Carroll FI (1994) Pharmaceutical Research 11:192
135. Bourguet W, Ruff M, Chambon P, Gonemeyer H, Moras D (1995) Nature 375:377

136. Zanotti G, D'Acunto MR, Malpeli G, Folli C, Berni R (1995) Eur J Biochem 234:563
137. Jaeger EP, Jurs PC, Stouch TR (1993) Eur J Med Chem 28:275
138. Krezel W, Ghyselinck N, Samad TA, Dupe V, Kastner P, Borelli E, Chambon P (1998) Science 279:863
139. Zanotti G, D'Acunto MR, Malpeli G, Folli C, Berni R (1995) Eur J Biochem 234:563
140. Thünemann A (1997) Langmuir 13:6040
141. Thünemann AF (2002) European Patent 1 003 559 and US Patent 6 395 284
142. Thünemann AF, Beyermann J, Ferber C, Löwen H (2000) Langmuir 16:850
143. Ponomarenko EA, Tirrell DA, MacKnight WJ (1998) Macromolecules 31:1584
144. Wenzel A, Antonietti M (1997) Adv Mater 9:487
145. Harada A, Cammas S, Kataoka K (1996) Macromolecules 29:6183
146. Bradbury E M, Crane-Robinson C, Goldman H, Rattle HWE (1998) Biopolymers 6:851
147. Botelho LH, Gurd FRN (1976) In: Fasman GD (ed) Handbook of Biochemistry and Molecular Biology, vol 2, 3rd edn. CRC Press: Cleveland, p 689
148. Lide DR (ed) Handbook of Chemistry and Physics, 73rd edn. CRC Press, Boca Raton, 1992
149. Stam CH (1972) Acta Crystallogr 28:2936
150. Douglas SJ, Davis SS (1985) J Colloid Interface Sci 103:154
151. Lechmann T, Reinhart WH (1998) Clin Hemorheol Microcirculat 18:31
152. Pons M, Garcia ML, Valls O (1991) Colloid Polym Sci 269:855
153. Thünemann AF, Beyermann J, Kukula H (2000) Macromolecules 16:5906
154. Allen C, Maysinger D, Eisenberg A (1999) Colloids and Surfaces B: Biointerfaces 16:3
155. Katayose S, Kataoka K (1996) In: Ogata N, Kim SW, Feijin J, Okando T (eds) Advanced Biomaterials in Biomedical Engeneering and Drug Delivery System. Springer, Berlin Heidelberg New York, p 319
156. Kataoka K, Togawa H, Harada A, Yasugi K, Matsumoto T, Katayose S (1996) Macromolecules 29:8556
157. Katayose S, Kataoka K (1997) Bioconjugate Chem 8:702
158. Harada A, Kataoka K (1995) Macromolecules 28:5294
159. Yokoyama M, Okano T, Sakurai Y, Ekimoto H, Shibazaki C, Kataoka K (1991) Cancer Res 51:3229
160. Kwon GS, Suwa S, Yokoyama M, Okano T, Sakurai Y Kataoka K (1994) J Controlled Release 29:17
161. Bronich TK, Nehls A, Eisenberg A, Kabanov VA, Kabanov AV (1999) Colloids and Surfaces B: Biointerfaces 16:243
162. Point JJ (1997) Macromolecules 30:1375
163. Han CH, Wiedmann TS (1998) International Journal of Pharmaceutics 172:241
164. Talibuddin S, Wu L, Runt J, Lin JS (1996) Macromolecules 29:7527
165. Dosiere M (1997) Macromol Symp 114:51
166. Shmueli U, Traub W (1965) J Mol Biol 12:205
167. Elliott A, Malcolm BR, Hanby WE (1957) Nature 179:960
168. Ponomarenko EA, Tirrell DA, MacKnight WJ (1996) Macromolecules 29:8751
169. Thünemann AF, Lochhaas KH (1999) Langmuir 15:4867
170. Antonietti M, Radloff D, Wiesner U, Spiess HW (1996) Macromolecular Chemistry and Physics 197:2713
171. Scheraga HA, Mattice WL (1987) Encycl Polym Sci Eng 7:685
172. Richardson JS, Richardson DC (1989) Prediction of Protein Structure and the Principles of Protein Conformation; Plenum Press: New York, pp 1-98
173. Ruokolainen J, Mäkinen R, Torkkeli M, Mäkelä T, Serimaa R, ten Brinke G, Ikkala O (1998) Science 280:557
174. Kataoka K, Ishihara A, Harada A Miyazaki H (1998) Macromolecules 31:6071
175. Greenfield N, Fasman GD (1969) Biochemistry 8:4108
176. Chou PY, Scheraga HA (1971) Biopolymers 10:657
177. Noy N (1992) Biochimia et Biophysica Acta 1106:159
178. Noy N (1992) Biochimia et Biophysica Acta 1106:151

179. Thünemann AF, Beyermann J (2000) Macromolecules 33:6878
180. Bougligand Y (1998) In: Handbook of Liquid Crystals, vol 1. Wiley-VCH, Weinheim, pp 409-410
181. Bronich TK, Cherry T, Vinogradov SV, Eisenberg A, Kabanov VA, Kabanov AV (1998) Langmuir 14:6101
182. Wolff T, Burger C, Ruland W (1994) Macromolecules 27:3301
183. Ruland W (1977) Colloid Polym Sci 255:417
184. Thünemann AF, Lochhaas KH (1998) Langmuir 14:6220
185. Holy V, Pietsch U, Baumbach T (1999) Springer Tracts in Modern Physics. High-resolution X-Ray Scattering from Thin Films and Multilayers, vol 149. Springer, Berlin Heidelberg New York
186. Rockley NL, Halley BA, Rockley MG, Nelson EC (1983) Analytical Biochemistry 133:314
187. Park IH, Choi E (1996) Polymer 37:313
188. Förster S, Hermsdorf N, Leube W, Schnablegger H, Regenbrecht M, Akari S, Lindner P, Bottcher C (1999) J Phys Chem B 103:6657
189. Borue V, Erukhimovich I (1990) Macromolecules 23:3625
190. Castelnovo M, Joanny JF (2001) Eur Physical J E 6:377
191. Castelnovo M, Joanny JF (2000) Langmuir 16:7524
192. Castelnovo M, Joanny JF (2002) Macromolecules 35:4531
193. Djadoun S, Goldberg R, Morawetz H (1977) Macromolecules 10:1015
194. Decher G (1997) Science 277:1232
195. Andelman D, Joanny JF (2000) Compt Rend Acad Sci IV 1:1153
196. Ladam G, Schaad P, Voegel JC, Schaaf P, Decher G, Cuisinier F (2000) Langmuir 16:1249
197. Borisov O, Zhulina E (1998) Eur Phys J E 4:205
198. Malthig B, Gohy JF, Jerome R, Stamm M (2001) J Polym Sci B 39:709
199. Hansen JP, Löwen H (2000) Annual Reviews of Physical Chemistry 51:209
200. Likos CN (2001) Physics Reports 348:267
201. Löwen H (1994) J Chem Phys 100:6738
202. Oosawa F (1971) Polyelectrolytes. Marcel Dekker, New York
203. Jusufi A, Likos CN, Löwen H (2002) Physic Rev Lett 88:018301
204. Jusufi A, Likos CN, Löwen H (2002) J Chem Phys 116:11011
205. Pincus P (1991) Macromolecules 24:2912
206. Likos CN, Hoffmann N, Jusufi A, Löwen H (2002) Interactions and Phase Behaviour of Polyelectrolyte Star Solutions (2002) J Phys: Condensed Matter (in press)

Received: November 2002

Polyelectrolyte Block Copolymer Micelles

Stephan Förster[1] · Volker Abetz[2] · Axel H. E. Müller[2]

[1] Institut für Physikalische Chemie, Universität Hamburg, Bundesstrasse 45, 20146 Hamburg, Germany
E-mail: forster@chemie.uni-hamburg.de
[2] Makromolekulare Chemie II and Bayreuther Zentrum für Kolloide und Grenzflächen, Universität Bayreuth, 95440 Bayreuth, Germany
E-mail: volker.abetz@uni-bayreuth.de
E-mail: axel.mueller@uni-bayreuth.de

Abstract Polyelectrolyte block copolymers form micelles and vesicles in aqueous solutions. Micelle formation and micellar structure depends on various parameters like block lengths, salt concentration, pH, and solvent quality. The synthesis and properties of more complicated block and micellar architectures such as triblock- and graft copolymers, Janus micelles, and core-shell cylinder brushes are reviewed as well. Investigations reveal details of the interactions of polyelectrolyte layers and electro-steric stabilization forces.

Keywords Block copolymers · ABC triblock copolymers · Janus micelles · Cylinder brushes · Core-shell nanoparticles · Graft copolymers · Micelles · Vesicles · Copolyampholytes · Polyelectrolyte block copolymers · Aggregation

1	Introduction	175
2	Micelle Formation of Polyelectrolyte Diblock Copolymers	175
2.1	Micellar Structure.	176
2.2	Salt Dependence.	179
2.2.1	Donnan Equilibrium.	181
2.3	Micellar Association Structures.	182
2.4	Electrosteric Interactions	187
2.5	Polyelectrolyte Vesicles	189
3	Triblock Copolymer Micelles and Vesicles.	191
4	Nonlinear Topologies and Nanoparticles	196
4.1	Janus Micelles	197
4.2	Core-Shell Cylinders	201
4.3	Graft Copolymers.	203
5	Conclusions and Outlook	206
	References	207

© Springer-Verlag Berlin Heidelberg 2004

Abbreviations

A	insoluble polymer block
B	soluble polymer block
b	grafting distance
c_s	salt concentration
l_B	Bijerrum length
N_{agg}	aggregation number
N_A	degree of polymerization of the soluble block
N_B	degree of polymerization of the insoluble block
N_L	Avogadros number
p_m	packing factor
R_c	micellar core radius
R_m	overall micellar radius
z	micellar charge
κ^{-1}	Debye length
ν	Flory exponent
P2VP	poly(2-vinylpyridine)
P4VP	poly(4-vinylpyridine)
PAA	poly(acrylic acid)
PAi	poly(*N,N*-dimethylaminoisoprene)
PB	poly(butadiene)
PCEVE	poly(chloroethyl vinylether)
PEE	poly(ethylethylene)
PEO	poly(ethyleneoxide)
PMAA	poly(methacrylic acid)
PMMA	poly(methyl methacrylate)
PtBS	poly(*tert*-butylstyrene)
PPO	poly(propyleneoxide)
PS	poly(styrene)
PnBA	poly(n-butylacrylate)
PSSH	poly(styrene sulfonic acid)
PSSNa	poly(Na-styrenesulfonate)
P2VPHCl	poly(2-vinypyridinium hydrochloride)
P2VPMeI	poly(*N*-methyl-2-vinylpyridinium iodide)
P4VPQ	quaternized poly(4-vinylpyridine)
DLS	dynamic light scattering
FFF	field flow fractionation
MALS	multi-angle light scattering
SANS	small-angle neutron scattering
SEM	scanning electron microscopy
SFM	scanning force microscopy
SLS	static light scattering
TEM	transmission electron microscopy

1
Introduction

Polyelectrolyte block copolymers combine structural features of polyelectrolytes, block copolymers, and surfactants. It is thus not surprising that they possess quite unusual and unique properties which make them a fascinating and challenging subject for researchers. Many of these properties are taken advantage of in technological applications and play an important role in physico-chemical properties of biological cell structures. This has motivated a comprehensive investigation so that today a much clearer picture of the behavior of polyelectrolyte block copolymers has developed.

The present review deals with the association behavior of polyelectrolyte block copolymers which is the most outstanding feature of this class of polymers. It leads to the formation of micelles, strings, and networks of sometimes quite complicated topology. The association behavior depends on many external parameters, among them pH, temperature, and salinity, which are of relevance in many technological and biological processes.

The following Sect. 2 deals with the association of diblock copolymers and discusses micelles, vesicles, micellar aggregation and interaction. In Sect. 3 recent work on micelle and vesicle formation of triblock copolymers is presented, whereas in Sect. 4 more complicated architectures such as graft copolymers, cylindrical brushes and Janus micelles are discussed.

2
Micelle Formation of Polyelectrolyte Diblock Copolymers

In dilute aqueous solutions, polyelectrolyte block copolymers self-assemble into micelles consisting of a hydrophobic core and a polyelectrolyte shell. The study of their structural properties is expected to provide a basic understanding of the properties of dense polyelectrolyte layers, electro-steric stabilization mechanisms, and actuator functions based on variations in the electrostatic interactions.

The first systematic investigation of micelle formation of polyelectrolyte block copolymers in aqueous solutions goes back to Selb and Gallot [1]. They noted the surprisingly low solubility of most of these polymers, when directly dissolved in water [1, 2]. Since the hydrophobic domains were in most cases glassy (e.g., in the case of polystyrene), thermal energy was insufficient for spontaneous dissolution. It then became common practice to use organic co-solvents (DMF, dioxane, THF) for the dissolution of the polymers, followed by dialysis to obtain stable, albeit "frozen" micelles in pure aqueous solutions [3]. Only in a few cases, was it possible to dissolve the polymers directly in water by using very short hydrophobic blocks [4] or by heating the solutions at 100 °C over a long period of time [5, 6]. "Frozen" micelles remain a major problem when investigating micellar properties since they do not correspond to the state of thermodynamic equilibrium. On the other hand, since their structure is "frozen", their micellar state can be kept constant when varying external parameters like salinity or temperature. Only in recent years, with the use of hydrophobic blocks with low glass transition temperatures ("soft blocks"), was it possible to investigate micellar properties under equilibrium conditions.

Systematic studies on micellar size and structure have been published for poly(styrene-*b*-acrylic acid) (PS-PAAc) [7, 8], poly(styrene-*b*-sodium acrylate) (PS-PAAcNa) [9], or quaternized poly(styrene-*b*-4-vinyl-pyridine) (PS-P4VPMeI) [10, 11]. It was concluded that the polyelectrolyte chains in the micellar corona are almost fully stretched [8]. The effect of salt concentration was investigated by Gucnoun et al. on poly(*t*-butylstyrene-*b*-sodium styrene sulfonate) (PtBS-PSSNa) who observed a weak decrease of micellar size and aggregation number when the salt concentration was increased beyond 0.01 mol/l [12]. Using small-angle neutron scattering (SANS), the authors could provide additional support for the rod-like conformation of the polyelectrolyte chains in the micellar corona [13].

2.1
Micellar Structure

Important parameters that control the size of micelles are the degree of polymerization of the polymer blocks, N_A and N_B, and the Flory-Huggins interaction parameter χ. The micellar structure is characterized by the core radius R_c, the overall radius R_m, and the distance b between adjacent blocks at the core/shell-interface as shown in Fig. 1. b is often called *grafting distance* for comparisons to polymer brush models. b^2 is the area per chain which compares to the area per head group in case of surfactant micelles. In the case of spherical micelles, the core radius R_c and the area per chain b^2 are directly related to the number of polymers per micelles, i.e., the aggregation number $Z=4\pi R_c^2/b^2$.

Aggregation numbers and core radii have been investigated in detail for many block copolymer/solvent systems [1, 14–18]. The core/shell structure

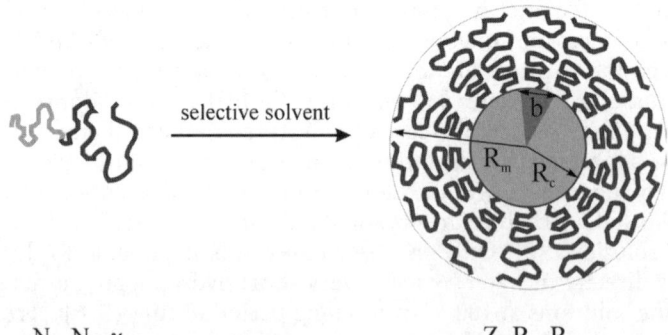

Fig. 1 Important parameters in the micellization of block copolymers. *A* is the insoluble block forming the micellar core, *B* is the soluble block forming the micellar shell or corona. Symbols are explained in the text

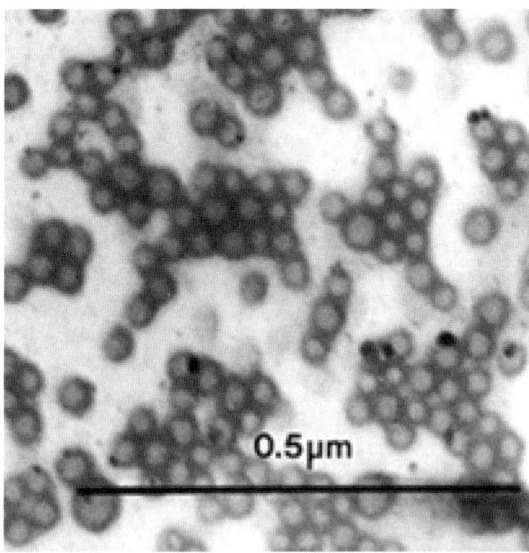

Fig. 2 TEM-image of spherical polyelectrolyte block copolymer micelles (PB-P2VP.MeI). The pronounced contrast of the polyelectrolyte shell is due to the counterions (I$^-$) [19]

can well be observed using transmission electron microscopy (TEM) [19] as shown in Fig. 2 in case of micelles formed by PB-P2VPMeI in water. In this case the core consists of the hydrophobic PB-blocks whereas the corona contains the polycationic P2VPMeI-blocks with the iodine counterions providing sufficient contrast for the TEM-image.

The free energy of a micelle is mainly determined by (1) the interfacial energy of the core/shell interface, and (2) the energy needed to stretch the block copolymer chains. The minimum of the free energy corresponds to an equilibrium grafting distance $b(N_A, N_B, \chi)$ which depends on block lengths and salt concentration (via χ). In the case of uncharged block copolymers, the grafting distance depends on the soluble block length as $b_0 N_B^{\beta/6}$ with $b_0 \approx 1$ nm. From this follows a simple relation for the aggregation number Z as a function of N_A and N_B

$$Z = Z_0 N_A^2 N_B^{-\beta} \qquad (1)$$

with $Z_0 \approx 1$ and $\beta \approx 0.8$ [20, 21]. As shown in Fig. 3 this equation successfully describes the micellization for uncharged diblock- [22, 44], triblock- [45], graft- [23], and heteroarmstar copolymers [24, 25] as well as for low molecular cationic, anionic, and non-ionic surfactants over three orders of magnitude in block length N_A.

The first theories on the micellization of polyelectrolyte block copolymers were published by Marko and Rabin [26], Dan and Tirrell [27] and Shusharina et al. [28]. They predict a strong influence of the polyelectrolyte blocks on the micellization behavior. Thus not unexpectedly, a qualitatively differ-

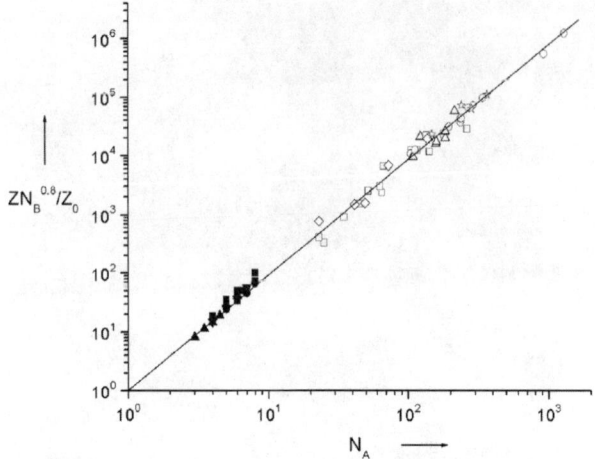

Fig. 3 Aggregation numbers Z as a function of the degree of polymerization of the insoluble block for uncharged block copolymers. *Open symbols* correspond to different diblock-, triblock-, graft- and star polymers. *Filled symbols* are low molecular weight surfactants [20, 21]

ent behavior is experimentally observed compared to uncharged block copolymers. The grafting distance shows a characteristic $b=b_0 N^{1/2}$-dependence as shown in Fig. 4 [29]. An increase in salt concentration lowers the value of R_m without affecting the 1/2-slope. This indicates that mutual repulsion of the polyelectrolyte chains leads to strong stretching of the core-blocks.

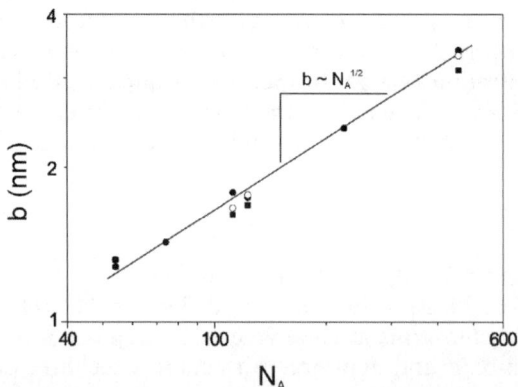

Fig. 4 Experimentally determined grafting distance b as a function of the degree of polymerization of the insoluble block for polyelectrolyte block copolymers at different salt concentrations [49]: (●) salt free, (○) 0.3 mol/l, (■) 1 mol/l

2.2
Salt Dependence

The addition of salt screens electrostatic interactions and thus has a strong influence on the conformation and interactions of polyelectrolyte chains [30, 31]. For polyelectrolyte block copolymer micelles, the effects are particularly pronounced because the polyelectrolyte chains are closely assembled in the micellar shell. When polymer chains are densely tethered to a surface to form a layer, they are described as a *polymer brush*. Theories for polyelectrolyte brushes have been developed using self-consistent field theories [32–36] and scaling theories [37, 38]. They describe how the brush thickness changes as a function of added salt concentration and grafting density b. It is found that depending on the added salt concentration, two regimes can be distinguished, in each of which the brush thickness shows a characteristic behavior. When the added salt concentration is larger than the concentration of counterions in the brush, the regime is called "*salted brush*". If on the other hand the added salt concentration is smaller than the brush counterion concentration, it is called "*osmotic brush*".

In the "*osmotic brush*" the osmotic pressure of the counterions leads to strong stretching of the polyelectrolyte chains. An increase in the added salt concentration or a decrease in grafting density has no effect on the brush height. The effect of added salt becomes considerable when the salt concentration in the bulk solution becomes comparable to or larger than the intrinsic ionic strength of the brush. In the respective "*salted brush*" regime the brush thickness D should decrease with increasing added salt concentration according to a $D \sim c^{-1/3}{}_s$-scaling law. These concepts have been used to describe the properties of polyelectrolyte block copolymers micelles.

The analogy to polyelectrolyte brushes was investigated by Guenoun in the study of the behavior of a free-standing black film drawn from a PtBS$_{26}$PSSNa$_{413}$-solution [39]. The thickness of these charged planar brushes as measured by X-ray reflectivity was found to decrease $D \sim c^{-1/3}{}_s$ when the salt concentration exceeded 0.2 mol/l. The analogy to charged brushes was further investigated by Harihanan et al. who studied the absorbed layer thickness of PtBS-PSSNa block copolymers onto latex particles [40, 41]. Similarly, the layer thickness was observed to correspond to the dimensions of fully stretched corona chains at low salt concentrations. When the salt concentration exceeded a certain limit, a weak decrease of the layer thickness with increasing salt concentration was observed. Similar results have been obtained by Tauer et al. [42] when investigating electrosterically stabilized latex particles. Recent experiments on charged planar brushes are in good agreement with theoretical predictions in the osmotic and salted brush regimes [43, 44].

The internal structure of polyelectrolyte block copolymer micelles such as their core radius R_c and micellar radius R_m can be determined by a variety of methods involving static and dynamic light scattering (SLS, DLS), small-angle X-ray (SAXS) and neutron scattering (SANS) as well as imaging techniques such as transmission electron microscopy (TEM) or atomic force mi-

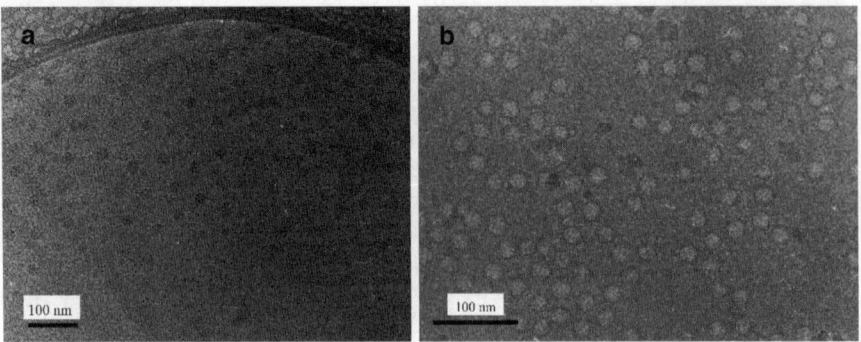

Fig. 5 Cryo-TEM images of polyelectrolyte block copolymer micelles (PEE-PSSH) at a NaCl-concentration of 0.003 mol/l (**a**) and 3 mol/l (**b**). The core/shell-structure is well visible in Fig. 5b [49]

croscopy (AFM) [45–49]. Using cryo-TEM one obtains direct images of polyelectrolyte micelles in their solvated state.

Figure 5 shows cryo-electron micrographs of micellar solutions of PEE-PSSH at two different salt concentrations c_s=0.003 (a) and 3 mol/l (b). Contrast arises from the electron density difference between micellar core, corona and the surrounding salt solution. In Fig. 5a (c_s=0.003 mol) the micellar cores appear as gray spherical domains. The radius of the micellar core is R_c=9.1 nm in good agreement with results from the SANS-experiments [49]. Addition of salt increases the electron density in the surrounding solution leading to contrast inversion. At a salt concentration of c_s=3 mol/l (Fig. 5b) the micellar cores appear as bright spherical domains surrounded by the dark interior micellar corona which contains a high concentration of ionic species ($-SO_3^+$, Na^+, Cl^-). The contrast directly corresponds to the spatial distribution of these ions. The measured radii are R_c=13.4 nm and R_m=26.3 nm which are also in good agreement with SANS. The images indicate a sharp, nearly discontinuous drop of the density at R_m.

From the measured core radii the grafting distance b can be calculated. Its variation as a function of added salt concentration is shown in Fig. 6 a. In the "*osmotic brush*" regime at low added salt concentrations the grafting distances are practically constant. At concentrations above c^*_s≈0.05 mol/l ("*salted brush*") the grafting distances decrease with increasing salt due to screening of the repulsive interactions between the corona chains [49].

According to theories in the *salted brush*-regime the layer thickness should decrease with increasing salt concentration. This is confirmed by experimental results shown in Fig. 6b, where above c^*_s≈0.05 mol/l the thickness decreases according to a $D \sim c_s^{-0.13}$-scaling law. This is in agreement with results from Guenoun et al. who reported a $D \sim c_s^{-0.14}$-dependence for PtBS$_{26}$PSSNa$_{404}$ and a $D \sim c_s^{-0.11}$-dependence for PtBS$_{27}$PSSNa$_{757}$ [12]. Adsorbed layers of PtBS$_{27}$PSSNa$_{757}$ on polystyrene latices exhibited a $D_h \sim c_s^{-0.17}$-dependence [41]. These values of the exponent are much smaller compared to theoretical predictions of $D \sim c_s^{-1/3}$. It is interesting to note in Fig. 6 that

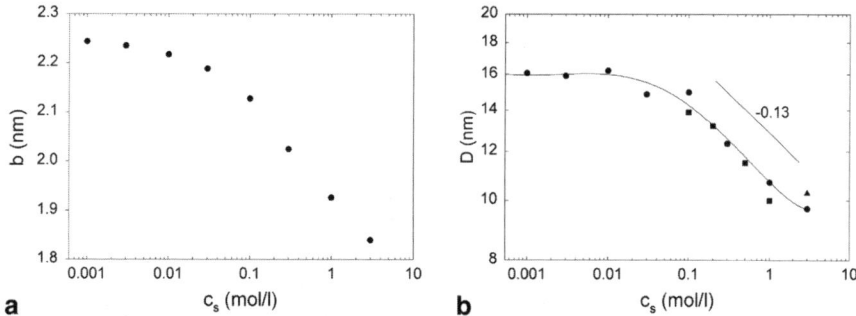

Fig. 6 Experimentally determined grafting distance b (**a**) and shell thickness D (**b**) as a function of NaCl-concentration for PEE-PSSH. Different behavior at low salt ("osmotic brush") and high salt concentration ("*salted brush*") can well be distinguished [49]

for a given salt concentration the layer thickness is similar for spherical micelles and planar brushes formed by the same polyelectrolyte block copolymer [43–45].

2.2.1
Donnan Equilibrium

From SANS-experiments it is possible to determine the ionic strength inside the polyelectrolyte shell, $\bar{c}_{s,\text{int}}$. Its dependence on the added salt concentration is shown in Fig. 7 [49].

At low ionic strength the local ion concentration is essentially constant and increases at larger added salt concentration. The c_s-dependence follows a trend that is characteristic for the Donnan equilibrium. As derived by Hariharan et al. [41] there should be a simple relation

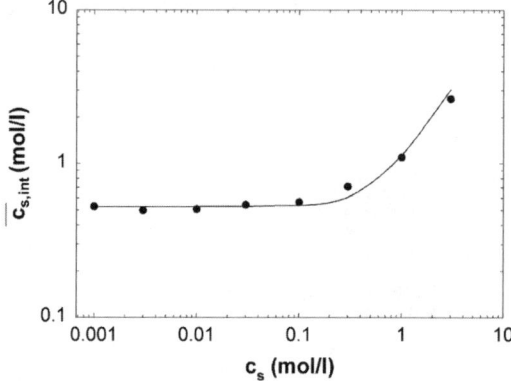

Fig. 7 Ionic strength of the micellar polyelectrolyte shell as a function of added salt concentration. The *solid line* is a fit to a simple Donnan equilibrium as described by Eq. (2) [49]

$$\bar{c}_{s,\text{int}} = c_s \left[1 + \frac{c_{s,0}^2}{c_s^2} \right]^{1/2} \quad (2)$$

between $\bar{c}_{s,\text{int}}$ and the added salt concentration c_s which allows to determine $c_{s,0}$, the intrinsic concentration of ionic species in the polyelectrolyte brush. From a fit of Eq. (2) to the data in Fig. 7, $c_{s,0}$ can be determined to be $c_{s,0}$=0.53 mol/l [49]. Experimentally, changes in micellar structure are observed above an added salt concentration of $c^*_s \approx$ 0.1 mol/l. This value is lower than the value of $c_{s,0}$ and indicates counterion condensation which reduces the effective ionic strength in the micellar corona. The Debye length in the corona, calculated from the interior salt concentration as $\kappa^2 = 8\pi N_L l_B c_{s,0}$ is $\kappa^{-1} \approx$ 0.4 nm. It is much smaller than the grafting distance b or the smallest blob size of the spherical brush and indicates that the local electroneutrality condition which requires $\kappa^{-1} \ll b$ is well fulfilled.

It was already noted in previous studies, that added salt affects the micellar structure only above a certain value of the salt concentration c^*_s. In studies of PtBS-PSSNa block copolymers a dependence of the brush dimension on the added salt concentration appeared above c_s=0.01 mol/l. The thickness of a free-standing black films drawn from a diblock polyelectrolyte solution exhibited a steady drop above ionic strengths of 0.2 mol/l [39].

2.3
Micellar Association Structures

It is usually assumed that the micellar corona is a continuous phase extending from the micellar core to the micellar radius R_m. The internal structure of the micelle can be described by a density profile as shown in Fig. 8. The micellar core is a homogeneous melt or glass of insoluble polymer blocks. For hydrophobic blocks in aqueous solutions, the polymer volume fraction in the micellar core is $\phi_c \approx$ 1. The micellar shell is swollen with water or aqueous salt solution and has a polymer segment density that is expected to decrease in the radial direction as $\phi(r) \sim r^{-\alpha}$ as typical for star polymers or

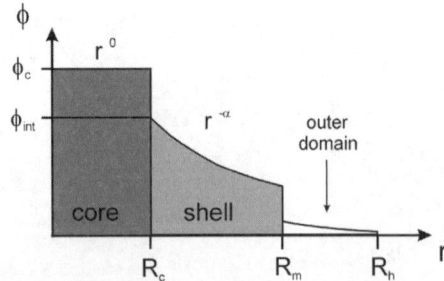

Fig. 8 Schematic density profile of a block copolymer micelle. Uncharged micelles exhibit a simple core/shell structure whereas polyelectrolyte block copolymer micelles can show phase separation of the corona into a dense interior and a dilute outer domain

spherical polymer brushes. At the core/shell interface there is a certain volume fraction of polymer chains, ϕ_{int} which is directly related to the interfacial distance b. Depending on the shell density at the outer periphery, the hydrodynamic radius is $R_h \geq R_m$.

For quite some time, there have been indications for a phase-separation in the shell of polyelectrolyte block copolymer micelles. Electrophoretic mobility measurements on PS-PMAc [50] indicated that a part of the shell exhibits a considerable higher ionic strength than the surrounding medium. This had been corroborated by fluorescence studies on PS-PMAc [51–53] and PS-P2VP-heteroarm star polymers [54]. According to the steady-state fluorescence and anisotropy decays of fluorophores attached to the ends of the PMAc-blocks, a certain fraction of the fluorophores (probably those on the blocks that were folded back to the core/shell interface) monitored a lower polarity of the environment. Their mobility was substantially restricted. It thus seemed as if the polyelectrolyte corona was phase separated into a dense interior part and a dilute outer part. Further experimental evidence for the existence of a dense interior corona domain has been found in an NMR/SANS-study on poly(methylmethacrylate-b-acrylic acid) (PMMA-PAAc) micelles [55].

The corona phase separation was predicted by Misra et al. [33] and Shusharina et al. [28] for charged brushes in poor solvents. Poor solvent conditions arise from the hydrophobic backbone of the polyelectroyte chain, for which water is a non-solvent. For charged brushes in poor solvents, phase separation should be most pronounced for spherical brushes because of the strong variation of segment density in the radial direction. According to de Gennes n-cluster theory, this phase separation is a consequence of a hierarchy of long-range repulsive forces (electrostatic), medium-range attractive forces (binary attraction of hydrophobic backbone) and short range repulsive forces (hard-core repulsion of backbone segments). As such the phase separation should be expected especially for small spherical polyelectrolyte brushes.

A detailed density profile for polyelectrolyte block copolymer micelles was obtained from a combination of SLS, DLS, and SANS together with cryo-TEM [49]. There is a phase separation in the corona with a dense interior part between $R_c < r < R_m$ ($R_c \approx 10$ nm, $R_m \approx 25$ nm) and a dilute outer part between $R_m < r < R_{out} \approx 40$ nm. The volume fractions in the interior part are in the range $0.06 < \phi < 0.35$, depending on salt concentration. The outer part of the corona has lower volume fractions $0.007 lt; \overline{\phi}_{out} lt; 0.04$. There is a sharp drop of the segment density at R_m which is evident in the cryo-electron microscopy images (Fig. 9b) and gives rise to neutron scattering at the corresponding interface.

From the scattering studies it is not possible to determine further details of the outer micellar part of the corona. Cryo-TEM studies provided further and rather surprising details [56]. In this study PB-P2VPHCl was investigated. Since the contrast is solely due to the counterions (Cl$^-$) the images represent a density map of the counterion distribution. Figure 10a shows a cryo-electron micrograph of isolated spherical micelles. The micellar cores appear as bright spherical domains due to the low electron density of the

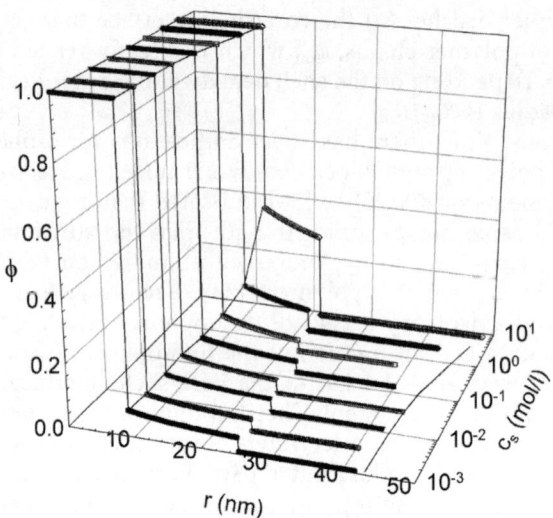

Fig. 9 Density profiles of PEE-PSSH micelles for varying salt concentrations. The profiles were obtained by a combination of static and dynamic light scattering, small-angle neutron scattering and cryo-TEM [49]

PB-core. The micellar core has a radius of R_c=17 nm in good agreement with scattering experiments. It is surrounded by a thin dark layer which represents the dense interior part of the polyelectrolyte corona containing a high concentration of condensed counterions.

In Figure 10b thin *filaments* are observed that extend from the interior shell into the surrounding solution, i.e. in the dilute outer part of the corona. The filaments have an average thickness of 2 nm and length of up to 50 nm which is close to the contour length of the polyelectrolyte chains (66 nm). They probably consist of polyelectrolyte chain bundles containing condensed counterions giving rise to the strong contrast observed in transmission electron microscopy. Where filaments merge with the interior layer, they have a characteristic catenoid shape. Catenoids are minimal surfaces with zero mean curvature and appear e.g. when pulling an elastic rubber membrane.

In Figure 10c the filaments connect adjacent micelles to form a *random filament network* with mesh sizes of 60 nm. Connected micelles form triangular or quadratic meshes that resemble the spectrin/actin network of erythrocyte cell membranes [57]. Micelles are completely separated and only connected by the filaments. Only in a few cases adjacent micelles are in direct contact with overlapping shells thereby forming doublets or triplets. Such clusters are also observed in Fig. 10d where micelles with overlapping shells are connected to form micellar strings that are surrounded by a single ion cloud or polyelectrolyte shell similar to electrons forming a molecular orbital around the atomic nuclei of a molecule.

There are weak attractive forces that lead to the formation of temporarily stable doublets, clusters and strings. Such forces are described by the theo-

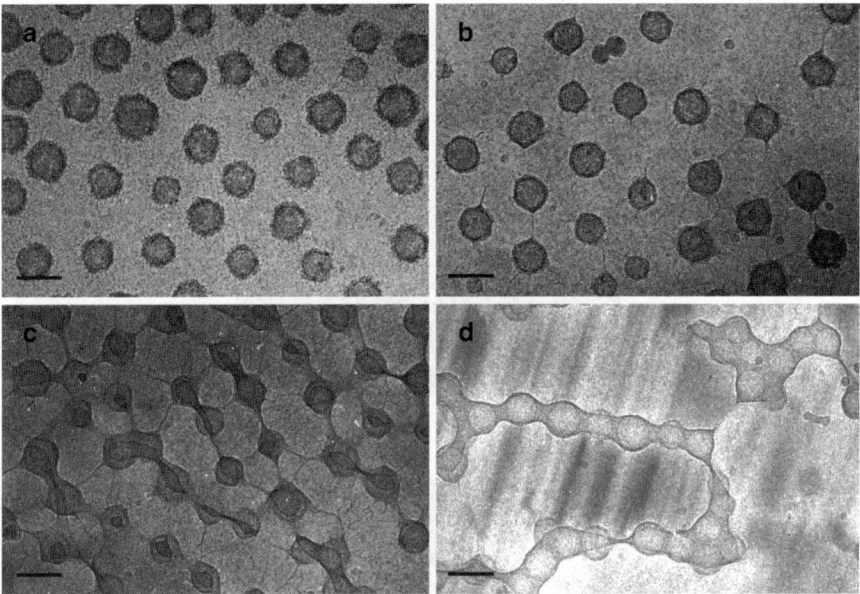

Fig. 10 Cryo-TEM images of polyelectrolyte block copolymer micelles (PB-P2VPMeI) with unperturbed spherical corona (**a**), corona filaments (**b**), filament networks (**c**), and micellar strings. The scale bar is 50 nm [56]

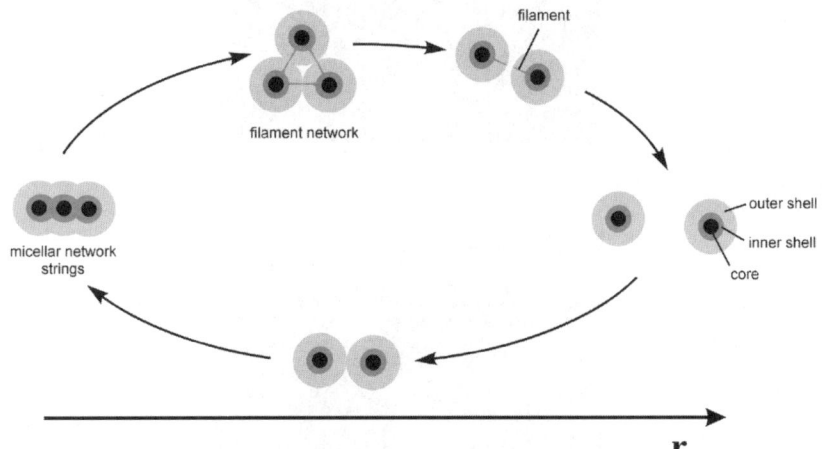

Fig. 11 Schematic view of a possible mechanism for the formation of micellar and filament networks. Different states of association are shown as function of intermicellar distance r

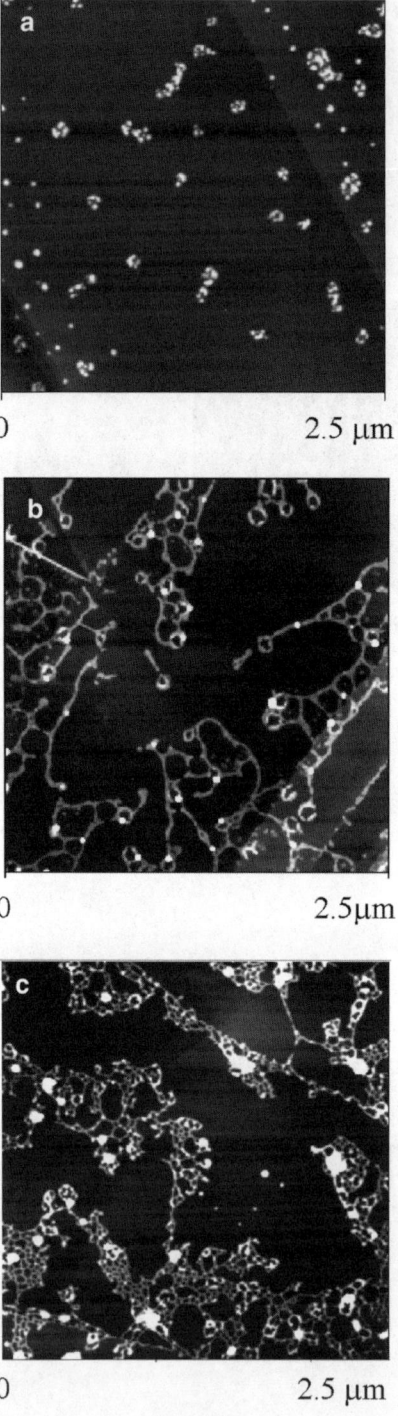

ries that predict phase separated corona structures. Accordingly, short-range attractive forces arise from transient fluctuations in the periphery of the corona, leading at times to attractive interactions between micelles. The effect of weak attractive interactions can be experimentally observed at intermediate added salt concentrations where the electrostatic repulsion between micelles is partially screened.

The formation of filaments, filament networks, and micellar strings is schematically shown in Fig. 11 as a function of the intermicellar distance r. Upon collisions small repulsive interactions are overcome to overlap adjacent micellar shells to form a shared ion cloud. The disruption of strings leads to the formation of filaments which can lead to a filament network. Further increase of intermicellar distance disrupts filament connections to form single micelles.

Using dynamic scanning force microscopy (SFM), it is possible to image the micellar strings and networks in more detail. SFM has become an efficient tool to investigate block copolymer micelles [58–60]. The micellar structures of PEE-PSSH are adsorbed from highly dilute solutions (50 mg/l) onto different substrates (graphite, mica) and are imaged with SFM in the Tapping Mode [45]. Aggregated states, i.e., strings and networks, can be induced by increasing the added salt concentration.

At no added salt there are spherical micelles (Fig. 12a) with an average micellar radius of 20 nm. Increasing the salt concentration to c_s=0.05 mol/l leads to the formation of micellar strings that exhibit a strong tendency to form crosslinks and loops or toroids (Fig. 12b). The toroids have diameters between 80 and 110 nm. At an added salt concentration of c_s=1 mol/l large networks are formed (Fig. 12c). The networks consist of connected toroids forming large loops with diameters covering all length scales similar to a fractal structure like the Sierpinski gasket. Further increase of salt concentration eventually leads to very large networks and macrophase separation into a dilute micellar phase and a concentrated gel phase [45, 46].

2.4
Electrosteric Interactions

Block copolymer micelles with their solvent swollen corona are a typical example of *soft spheres* having a soft repulsive potential [61]. The potential has been derived by Witten und Pincus for star polymers [62] and is of form $u(r)\sim\ln(r)$. It only logarithmically depends on the distance r and is therefore much softer compared to common r^{-x}-potentials such as the Lennard-Jones potential (x=12). The potential is given by

Fig. 12 AFM-images of polyelectrolyte block copolymer micelles (PEE-PSSH) at no added salt (**a**), 0.1 mol/l (**b**) and 1 mol/l (**c**). The formation of strings, loops and networks occurs with increasing salt concentration [45, 46]

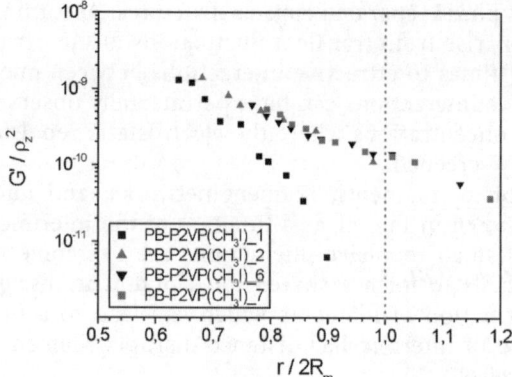

Fig. 13 Measured shear modulus G' as a function of the reduced intermicellar distance. at $r/2R_m<1$ the shells of adjacent micelles start to overlap. The linear behavior in the semi-logarithmic presentation is characteristic for a simple Debye-Hückel type interaction potential [63]

$$u_{\text{steric}}(r) \cong vkTZ^{3/2}\ln\left(\frac{r}{2R_m}\right) \Leftrightarrow G'(r) \cong p_m vkTZ^{3/2}\frac{1}{r^2} \quad (3)$$

where p_m is a packing factor and z the micellar charge. It is experimentally difficult to directly determine the interaction potential $u(r)$. However, it is straightforward to determine its second derivative, i.e., the shear modulus $G'(r)$ as a function of distance for micellar gels. For uncharged block copolymer micelles the predicted $G'(r) \sim r^{-2}$ behavior is experimentally observed and there is nearly quantitative agreement between theory and experiment [61]. The potential $u(r)$ corresponds to the *steric interaction* potential of polymer- or sterically stabilized colloids.

Since neutral block copolymer micelles are a convenient model system for the investigation of the steric stabilization potential, polyelectrolyte block copolymer micelles may serve to study *electro-steric interaction*. Similar to the case of neutral block copolymer micelles, the interaction potential $u(r)$ can be probed by measuring the shear modulus of micellar gels as a function of distance. Assuming a simple Debye-Hückel potential yields for the shear modulus by taking the second derivative

$$u_{\text{electro}}(r) \cong z^2 l_B kT \frac{e^{-\kappa r}}{r} \Leftrightarrow G'(r) \cong p_m z^2 l_B kT\left(2 + 2\kappa r + \kappa^2 r^2\right)\frac{e^{-\kappa r}}{r^3} \quad (4)$$

This is the dependence expected for purely electrostatic stabilization of polyelectrolyte layers. *Electro-steric interactions* involve steric contributions [Eq. (3)] and electrostatic contributions [Eq. (4)]. For high ionic strength as in the interior of the polyelectrolyte shell the measured shear modulus should exhibit a characteristic $G'(r) \sim e^{-\kappa r}$-dependence which is apparent in a semi-logarithmic presentation of the data as in Fig. 13. The measured shear moduli are plotted as a function of the reduced distance $r/2R_m$ which is equal to 1 at contact of adjacent micellar shells [63]. Electrostatic interactions are approximately one order of magnitude larger compared to purely steric interactions. Sample #1 has only very weak repulsive forces at the periphery indicating very weak electro-steric stabilization. If the interactions are further weakened, e.g., by the addition of salt, this leads to weak aggregation into micellar clusters and strings as outlined in Sect. 2.3.

2.5
Polyelectrolyte Vesicles

Spherical micelles are not the only association structure that is formed by polyelectrolyte block copolymers. With increasing hydrophobic block length there is a tendency to form block copolymer vesicles. A vesicle formed by PB-P2VP.HCl is shown in the cryo-TEM image in Fig. 14a. The bilayer structure is clearly resolved which shows that block copolymer vesicles are structurally very similar to lipid vesicles. Vesicles can be also imaged by AFM (Fig. 14b) where they exhibit a characteristic outer rim because the interior solution of the vesicle has evaporated during sample preparation leaving a shape that resembles that of an empty football. Vesicles typically have diameters of 100–300 nm and a bilayer thickness of 10–20 nm.

In the cryo-TEM image (Fig. 14a), vesicles have a grainy bilayer surface due to a large number of filaments. They are surrounded by shell of micelles and filament networks. There is a coexistence of spherical micelles and vesicles over a wide concentration range from very dilute to concentrated, gel-like solutions. By using a sol-gel templating technique, thin sections of these gels can be microtomed and imaged by TEM. With this method it is possible to obtain a clearer picture of the vesicle interior. As seen in Fig. 14c, many vesicles contain spherical micelles. In some cases onion-type multilamellar vesicles (liposomes) are formed [64].

Also vesicles of P2VP-PEO block copolymers can be prepared in aqueous solutions [65]. It is possible to form giant vesicles with diameters of 5–10 μm which can be imaged with optical microscopy. An interesting property of these vesicles is their pH-dependent stability. Vesicles are stable down to pH 4.5, below which the P2VP-block is protonated and the vesicles dissolve. The dissolution process can be followed with video microscopy as shown in Fig. 15. A front of dilute acetic acid with pH 4 is moving in from the right hand to the left hand side. Within seconds the vesicles rupture and completely dissolve. This is of relevance in the chemotherapy of cancer cells which due to their high metabolism have slightly acidic cytosols. If such vesicles are used as drug carriers, they only release the drug inside these cells.

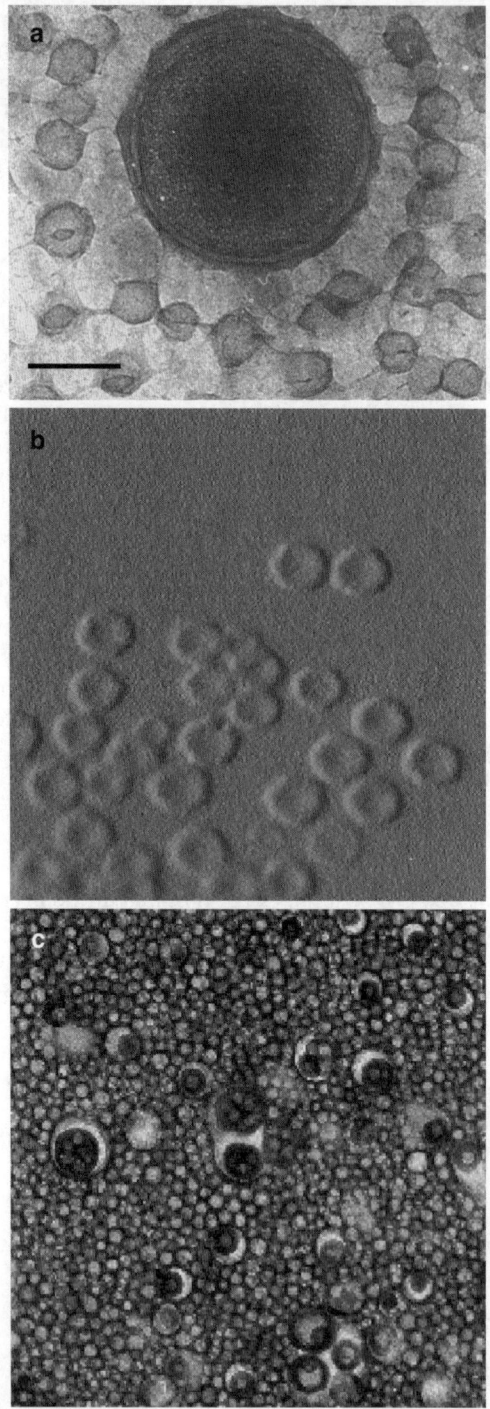

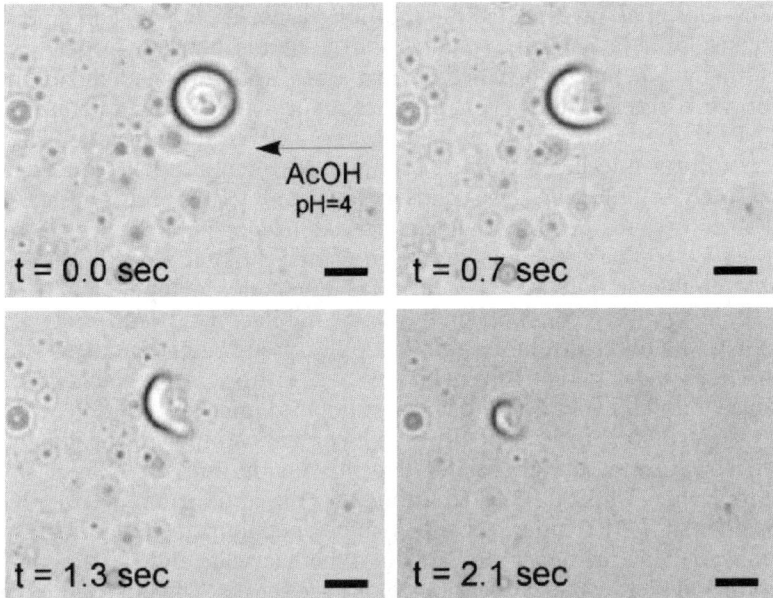

Fig. 15 Video microscopy snapshots of the rupture and dissolution of block copolymer vesicles (P2VP-PEO) upon addition of acetic acid. Such vesicles may be used for the pH-controlled delivery of drugs [65]

3
Triblock Copolymer Micelles and Vesicles

So far, micelles and vesicles of amphiphilic block copolymers with two different blocks have been described. In this section the work on amphiphilic block copolymers and block copolyampholytes composed of three different blocks will be reviewed. Much less work has been carried out on these systems and there are less systematic studies available. Focus will be laid on block copolymers with at least one polyelectrolyte block. While in the case of amphiphilic diblock copolymers questions like the influence of block lengths on the size of micellar aggregates have been studied in great detail, in ternary block copolyampholytes other properties have attracted greater interest, such as the influence of the block sequence on the solution properties and aggregate formation.

Fig. 14 Cryo-TEM (a), AFM- (b), and TEM images of polyelectrolyte block copolymer vesicles (PB-P2VP.MeI). The image (c) is taken from a silica-template which was obtained by a sol/gel-process of a concentrated micellar solution [47, 56, 64]

Patrickios et al. investigated the solution properties of a system composed of an insoluble, a partially soluble, and a soluble block in water, namely poly(ethyl vinyl ether), poly(methyl vinyl ether) and poly(methyl tri(ethylene glycol) vinyl ether) with varying block sequences [66]. While in aqueous solutions only unimers were found, addition of salt led to aggregates. The tendency to form micelles was most pronounced when a block sequence with increasingly soluble blocks was chosen. Kriz et al. [67] found spherical core-shell structures for poly(2-ethylhexyl acrylate)-*block*-poly(methyl methacrylate)-*block*-poly(acrylic acid) in water. While in that system the water-soluble endblock poly(acrylic acid) was the major component, Yu et al. [68] studied polystyrene-*block*-poly(methyl methacrylate)-*block*-poly(acrylic acid) with the water insoluble endblock polystyrene being the major component. They used different solvents (dioxane, THF, or DMF) which were continuously diluted by addition of water and found micelles of different shapes and vesicles as a function of the water content. Moreover, the finally obtained structure of the aggregates depends on the initially chosen solvent mixture, which indicates the freezing of non-equilibrium morphologies. Ishizone et al. [69] synthesized ABC triblock copolymers containing 2-(perfluorobutyl)ethyl methacrylate, *tert*-butyl methacrylate and 2-(trimethylsilyloxy)ethyl methacrylate with various block sequences. The block copolymers then were converted into amphiphilic systems by removing the trimethylsilyl protecting group to give a poly(2-hydroxyethyl methacrylate) block. ^{1}H-NMR spectra of the resulting triblock copolymers suggested the formation of their micelles in selective solvents.

Amphiphilic block copolyampholytes in which two of the blocks possess opposite charges besides a third, hydrophobic block constitute a small and scarcely investigated but highly interesting class within the amphiphilic block copolymer family. Patrickios et al. [70, 71, 72] studied ABC triblock copolyampholytes consisting of 2-(dimethylamino)ethyl methacrylate, methyl methacrylate or 2-phenylethyl methacrylate, respectively, and methacrylic acid with similar molar fractions. Rather low molecular weight materials were studied and no influence of the block sequence on the isoelectric point was found. However, depending on the block sequence, different sizes were found for the micellar superstructures in aqueous solution. The largest micelles were found for the two systems with a hydrophobic poly(methyl methacrylate) end block, while the block copolymer with a centered PMMA block separating the polycationic and poly anionic blocks displayed a smaller diameter.

Giebeler et al. [73] investigated the polyelectrolyte complex formation of triblock copolyampholytes, polystyrene-*block*-poly(2 or 4)-vinylpyridine)-*block*-poly(methacrylic acid). By potentiometric, conductometric and turbidimetric titrations of acidic THF/water solutions the formation of an interpolymer complex at the isoelectric point was found, in which most likely the hydrophobic polystyrene cores are embedded in a mixed corona of the two polyelectrolyte blocks.

Bieringer et al. [74] studied poly(5-(N,N-dimethylamino)isoprene)-block-polystyrene-block-poly(methacrylic acid) (AiSA) triblock copolyampholytes. In this system the first block forms a polycation at low pH, while the last block forms a polyanion at large pH. The center block remains always neutral. Investigations of the pH-dependent solution properties of the AiSA triblock copolyampholytes in solution had to be performed in solvent mixtures due to a complex solubility behavior of these materials. The hydrochloride of poly(5-(N,N-dimethylamino)isoprene) and the sodium salt of poly(methacrylic acid) are only soluble in water; poly(methacrylic acid) is best soluble in lower alcohols whereas poly(5-(N,N-dimethylamino)isoprene) as well as polystyrene dissolve in organic solvents like THF, toluene, chloroform, or cyclohexane. For the investigation of the AiSA solution behavior THF/water mixtures proved most valuable. The rather viscous solutions indicated the formation of aggregates. These systems could not be dissolved in pure water, even dialysis of THF/water solutions failed and lead to precipitation.

The different situations of the block copolyampholyte as a function of pH are depicted in Scheme. 1. The same order of deprotonation was observed in copolymers of poly(methacrylic acid) and poly(2-(diethylamino)ethyl methacrylate) [70, 71, 72, 75–77]. In copolymers of poly(methacrylic acid) [78] and poly(2/4-vinylpyridine), on the other hand, deprotonation of the pyridine hydrochloride takes place prior to deprotonation of the carboxyl [79–83] because in comparison to carboxylic functionalities amine hydrochlorides are the weaker acids and vinylpyridinium hydrochlorides the stronger ones. Thus, in the latter systems an isoelectric point (iep) is not observable, while in the first case polyzwitterions are formed which possess the highest amount of charges at that pH resulting in a polyelectrolyte complex (PEC). In the second case polymers with a minimal net charge built the PEC which is stabilized by hydrophobic interactions often accompanied by hydrogen bonding [79, 84–88].

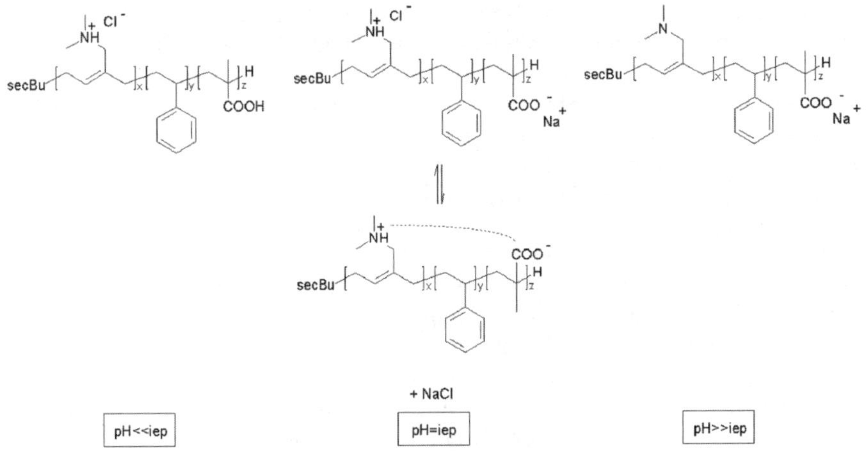

Scheme1 pH Dependence of the functional groups of the triblock copolyampholytes [74]

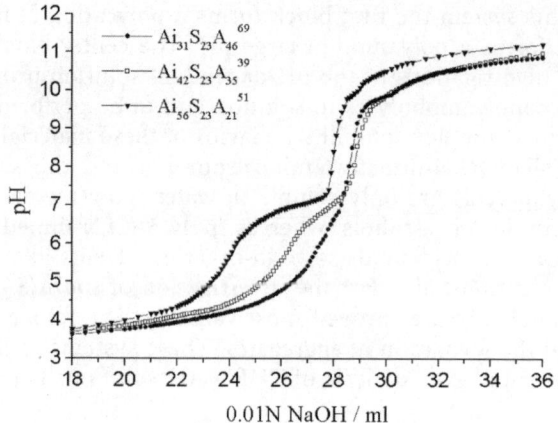

Fig. 16 Potentiometric titrations of three different triblock copolyampholytes [74]

As the charge density on the polymer backbone increases during the titration of poly(methacrylic acid) with sodium hydroxide, deprotonation becomes more difficult in the course of titration and the pKa value increases [89–91].

Potentiometric titration curves of AiSA triblock copolyampholytes in THF/water (2:5 weight/weight) are shown in Fig. 16 for three copolymers possessing identical mole fractions of hydrophobic polystyrene but different amounts of acidic and basic end blocks.

Independent of the composition all triblock copolyampholytes displayed two inflection points at pH≈5.5 and 9 which correspond to the deprotonation of the acid and amine hydrochloride functionalities, respectively.

Dynamic light scattering of an AiSA triblock copolyampholyte in THF/water (2:5 weight/weight) solution showed that essentially one defined particle type with low polydispersity can be assumed. The hydrodynamic radius of this species is found to vary with pH and ranges from 260 to 120 nm showing that highly aggregated structures were obtained. In Fig. 17 the titration curve of $Ai_{57}S_{11}A_{32}$ [37] together with the hydrodynamic radii at different pH values is shown. The indices denote the weight fractions whereas the superscript is the molecular weight in kg/mol.

The decrease in the hydrodynamic radius coincides with the first inflection point of the potentiometric titration; a further and significantly smaller decrease might be assumed at the second inflection point. At pH 4 spherical structures with a maximal radius of approximately 120 nm are observed. As the contour length of the polymer chain is only 140 nm these spherical aggregates most probably do not represent simple spherical micelles. More likely vesicles are formed. At pH 10 also spherical aggregates are found, which appear to be less uniform than the ones formed under acidic conditions. at basic conditions the maximal radius was determined to be about 100 nm.

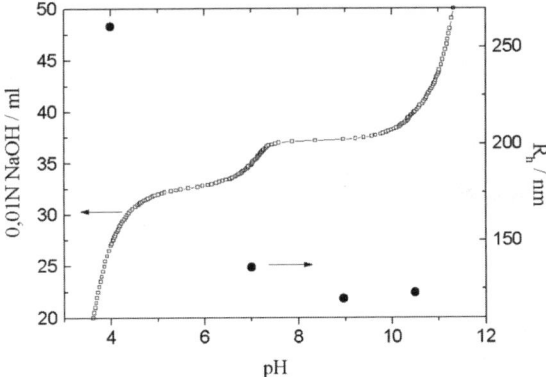

Fig. 17 Titration curve of $Ai_{57}S_{11}A_{32}$ [37] and dependence of the hydrodynamic radii on pH [74]

Suggestions of possible vesicular structures under the two pH conditions are shown in Fig. 18. Under acidic conditions both end blocks are soluble and the insoluble shell is formed by the polystyrene block. On the contrary, at basic conditions both polystyrene and poly(5-(N,N-dimethylamino)isoprene) are insoluble and only the deprotonated poly(methacrylic acid) block keeps the vesicular structure in solution.

As the solutions at pH other than 4 were prepared from the corresponding acidic solutions by adding sodium hydroxide solution until the desired pH was reached it is not surprising that the vesicle dimensions are not found to depend on the pH of the solution to a great extent. The glassy polystyrene center-block defines the pH-independent diameter of the inner vesicle shell, i.e., the structure is frozen in terms of the aggregation number of involved chains (frozen vesicles). at higher pH values the PAi precipitates on the PS shell (Fig. 18b). The only way the diameter of the vesicle can vary with pH is by a change in the conformation of the soluble end block(s) due to changing of the charge density.

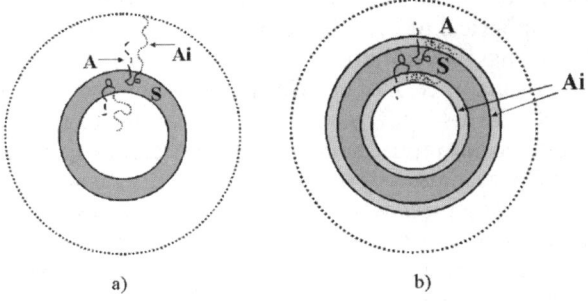

Fig. 18 Possible vesicular structures at pH 4 (a) and 10 (b) [74]

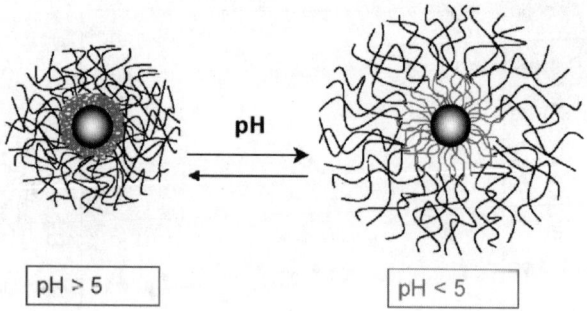

Fig. 19 Effect of pH on the size of the micelles (PS core as a sphere, P2VP chains in *blue*, PEO chains in *black*) [92]

Gohy et al. studied the solution properties of micelles formed by two polystyrene-*block*-poly(2-vinylpyridine)-block-poly(ethylene oxide) (PS-*b*-P2VP-*b*-PEO) copolymers in water by dynamic light scattering and transmission electron microscopy [92]. Spherical micelles were observed that consist of a PS core, a P2VP shell and a PEO corona. The characteristic sizes of core, shell and corona were found to depend on the copolymer composition. The micellar size increased at pH<5 due to P2VP block protonation (Fig. 19).

This change was found to be reversible if the corona block is not too large and may become of interest for controlled drug delivery applications. It was also shown that these systems can host gold nanoparticles in the P2VP shell [93].

4
Nonlinear Topologies and Nanoparticles

Polymeric colloids or nanoparticles are an interesting class of materials. Besides microgels of crosslinked homopolymers, also other chain topologies can be used to generate nanoparticles. For example, block copolymers can act as precursors of nanoparticles, spherical or cylindrical micelles formed by block copolymers in solution can be crosslinked in the shell or in the core, leading to spherical or cylindrical nanoparticles. However, only the remaining polymer has flexible polymer chains. Crosslinking of the B block of a core-shell-corona micelle of an ABC triblock copolymer is a way to stabilize structures with two different kinds of polymer chains which can be further modified. For example, hollow structures can be generated in a subsequent selective degradation of the core (A) domain, as was shown by Liu's group [94]. These structures were discussed also in terms of their potential as compartments for guest molecules, like drugs or dyes. An interesting alternative to these centrosymmetric structures are non-centrosymmetric, surface-compartmentalized micelles, in which a core is surrounded by two different corona hemispheres, so-called Janus micelles. These will be reviewed in Sect. 4.1.

Starblock (or radial star) copolymers form another kind of amphiphilic nanoparticles which can be regarded as unimolecular micelles. Alternatively, cylindrical core-shell brushes can be regarded as unimolecular cylinder micelles. Due to the covalent attachment of the block copolymers at one end, "frustrated" micellar structures can be made which would never form spontaneously. The cylindrical systems will be reviewed in Sect. 4.2.

Finally, amphiphilic graft copolymers will be reviewed and compared to the linear block copolymer structures of the same composition and it will be shown that micellization depends not only on composition, pH, and ionic strength, but also significantly on the topology of the polymer molecule.

4.1
Janus Micelles

Janus micelles are non-centrosymmetric, surface-compartmentalized nanoparticles, in which a cross-linked core is surrounded by two different corona hemispheres. Their intrinsic amphiphilicity leads to the collapse of one hemisphere in a selective solvent, followed by self-assembly into higher ordered superstructures. Recently, the synthesis of such structures was achieved by crosslinking of the center block of ABC triblock copolymers in the bulk state, using a morphology where the B block forms spheres between lamellae of the A and C blocks [95, 96]. In solution, Janus micelles with polystyrene (PS) and poly(methyl methacrylate) (PMMA) half-coronas around a crosslinked polybutadiene (PB) core aggregate to larger entities with a sharp size distribution, which can be considered as supermicelles (Fig. 20). They coexist with single Janus micelles (unimers) both in THF solution and on silicon and water surfaces [95, 97].

By hydrolysis of the ester groups of the PMMA arms, an amphiphilic polyelectrolyte was obtained with PS arms and poly(methacrylic acid) (PMAA) arms. [98, 99] These particles are soluble in water, when using several sequential dialysis steps (starting from 1,4-dioxane). However, due to the non-polarity of both PB core and PS hemi-corona they collapse to a core

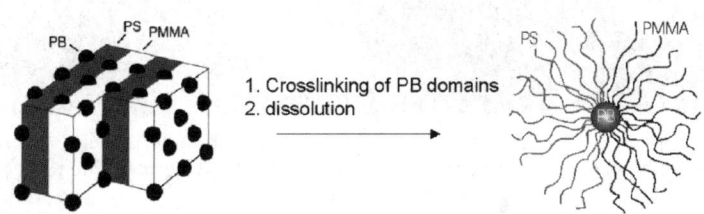

Fig. 20 Scheme of the synthesis of a PS-PB-PMMA Janus micelle. Reprinted with permission from ref [95]. Copyright (2001) American Chemical Society

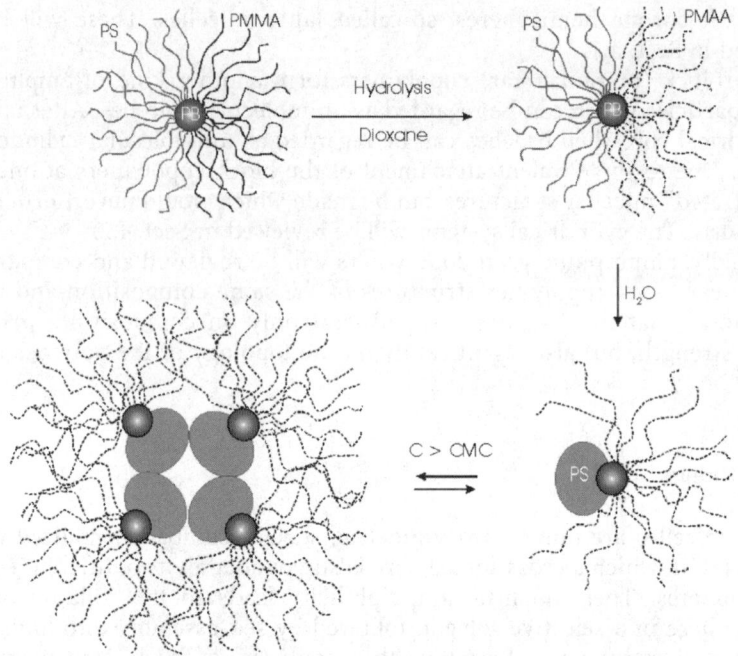

Fig. 21 Synthesis and tentative structure of amphiphilic Janus micelles and their supermicelles. Reprinted with permission from ref [99]. Copyright (2003) American Chemical Society

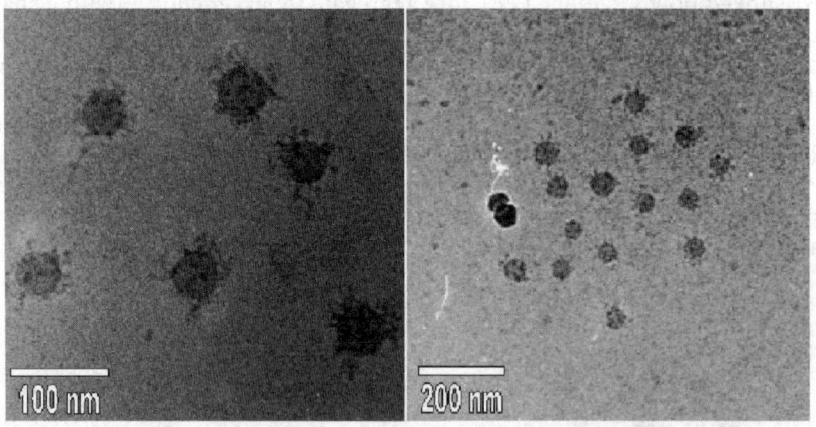

Fig. 22 Cryogenic transmission electron micrograph of supermicelles in solution. Reprinted with permission from ref [99]. Copyright (2003) American Chemical Society

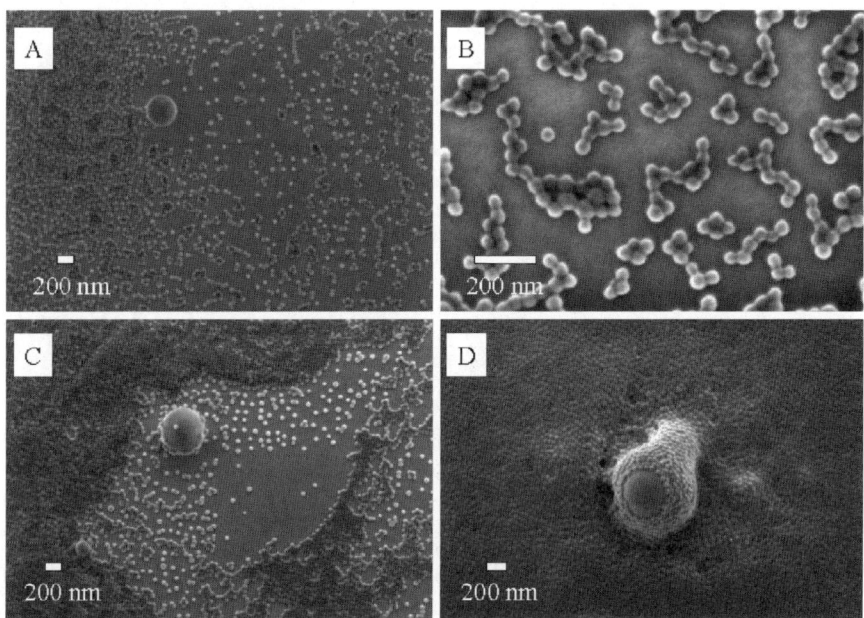

Fig. 23 Scanning electron micrograph of supermicelles deposited on a silicon wafer. Reprinted with permission from ref [99]. Copyright (2003) American Chemical Society

which is fully or partially surrounded by the swollen PMAA arms. The vast majority of these particles self-assembles to larger, spherical entities of well-defined size (and thus aggregation number). Most likely, the non-polar components form the core and the PMAA arms are stretched into the surrounding solution. This situation is comparable to classical surfactant micelles in which the building entities are replaced by polar spheres, as schematized in Fig. 21.

These structures were proven by various methods. Cryogenic transmission electron microscopy confirmed the existence of the supermicelles in solution (Fig. 22), while scanning electron microscopy (Fig. 23) and scanning force microscopy could prove larger particles with a non-uniform size of still unknown nature besides the supermicelles and unimers.

Static and dynamic light scattering as well as asymmetric flow field-flow fractionation (AF-FFF) indicate the presence of supermicelles and of significantly larger structures in aqueous solution. Figure 24 shows the results of AF-FFF, coupled with a multi-angle light scattering (MALS) detector. The eluogram clearly shows two distinguishable species, especially at low scattering angles. The very strong angular dependence for the second species indicates its large size, whereas its amount is so small that it is almost invisible in the concentration signal. The radii of gyration of two species were determined to be R_g=63 nm and 118 nm, respectively, and from the molecular weight of the first peak a number of 37 Janus micelles per particle was calculated.

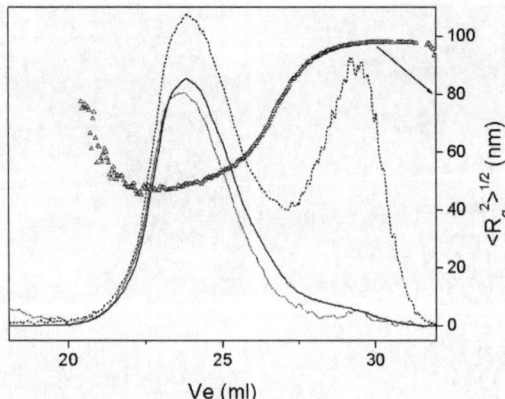

Fig. 24 AFFFF-MALS measurements of Janus micelles in water with 1 wt-% NaCl. (—): LS 90°;(- - -): LS 35°; (·····): RI; (Δ):radius of gyration. Reprinted with permission from ref [99]. Copyright (2003) American Chemical Society

The pH-dependence of the hydrodynamic radii of both supermicelles and the much larger entities is shown in Fig. 25. For acidic conditions the hydrodynamic radius of the supermicelles is smaller than for basic conditions, which is due to the ionization of the PMAA arms leading to their stretching.

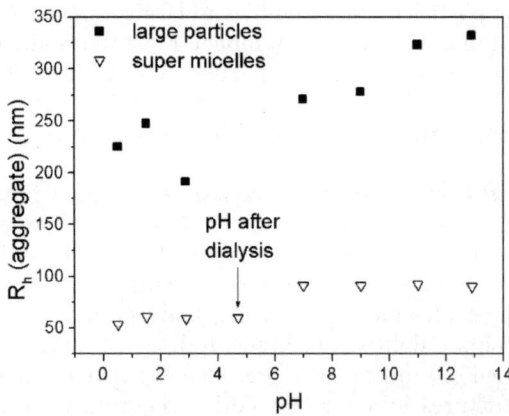

Fig. 25 Hydrodynamic radius of supermicelles in water with 1 wt-% NaCl as a function of pH. Reprinted with permission from ref [99]. Copyright (2003) American Chemical Society

4.2
Core-Shell Cylinders

Cylindrical wormlike micelles have been investigated by many groups in recent years [100] most of them being formed by aggregation of surfactants. As an example, cetyltrimethylammonium bromide reversibly assembles into long, flexible wormlike micelles in 0.1 M KBr aqueous solution. These aggregates may dissociate or undergo structural changes under changed conditions. Similarly, block copolymers can form spherical or cylindrical micelles in selective solvents [101, 102, 103]. Although spheres are the most common morphology for block copolymer micelles, other types of supramolecular structures such as cylinders have also been found. For example, polyferrocenylsilane-*block*-poly[2-(*N,N*-dimethylamino)ethyl methacrylate] with a block ratio of 1:5 formed cylindrical micelles in aqueous solution [104]. The cylindrical structures can be stabilized in a covalent way by synthesizing core-shell cylinder brushes, moreover, core-shell structures can be made which would never form by self-aggregation and their length can be much better controlled.

Generally, there are three methods to synthesize cylindrical polymer brushes. The first one, which was widely used in the past decade, is the conventional radical polymerization of macromonomers [105–111]. However, these polymerizations normally yield polymer brushes with a broad chain-length distribution. Aiming at better-defined brushes, living anionic [112] and ring-opening metathesis polymerization (ROMP) [113–115] of macromonomers were also performed, however, high molecular weight polymers have not been prepared, so far. The second method is the "grafting onto" technique [116–118]. For example, coupling living polystyryllithium chains with poly(chloroethyl vinyl ether) (PCEVE) resulted in a polymer brush with PCEVE as backbone and polystyrene (PS) as side chains [116, 117]. However grafting efficiency was often insufficient. Finally, in the "grafting from" process, the side chains of the brush are formed via atom transfer radical polymerization (ATRP) [119], initiated by the pendant initiating groups on the backbone [120–122. By this method well-defined polymer brushes with high grafting density and rather narrow distributions of both backbone and side chains can be obtained, and the purification of resulting polymer brushes is much simpler comparing to the other two methods.

So far, there have been only few reports about the synthesis of amphipolar polymer brushes, i.e. with amphiphilic block copolymer side chains. Gnanou et al. [115] first reported the ROMP of norbornenoyl-endfunctionalized polystyrene-*b*-poly(ethylene oxide) macromonomers. Due to the low degree of polymerization, the polymacromonomer adopted a star-like rather than a cylindrical shape. Schmidt et al. [123] synthesized amphipolar cylindrical brushes with poly(2-vinylpyridine)-*block*-polystyrene side chains via radical polymerization of the corresponding block macromonomer. A similar polymer brush with poly(α-methylstyrene)-*block*-poly(2-vinylpyridine) side chains was also synthesized by Ishizu et al. via radical polymerization [124]. Using the "grafting from" approach, Müller et al. [121, 125] synthesized

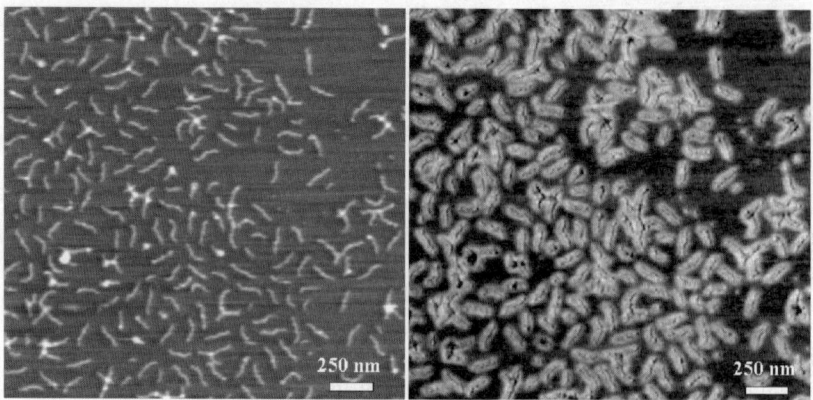

Fig. 26 SFM Tapping Mode images of the brush, $[(AA)_{31}\text{-}b\text{-}(nBA)_{48}]_{1500}$, dip-coated from dilute $CH_3OH/CHCl_3$ (1/1) solution on mica: (*left*) height image (z-range: 6 nm) and (*right*) phase image (range: 40 °). Reprinted with permission from ref [99]. Copyright (2003), with the permission from Elsevier Science

core-shell cylindrical brushes with polystyrene (PS) or poly(n-butyl acrylate) (PnBA) core and poly(acrylic acid) (PAA) shell as well as the inverse unimolecular micelles.

The core-shell cylinder brushes can be visualized by scanning force microscopy (SFM). Figure 26 shows cylinders with 1500 side chains with 31 AA and 48 nBA units each, which have a very uniform size distribution [125].

The ability of the hydrophilic PAA core of the amphiphilic core-shell brushes to coordinate with different metal cations can be used for the synthesis of novel nanosized organic/inorganic hybrids or for the generation of gold clusters or cobalt nanowire [126].

The unimolecular cylinder micelles undergo an anisotropic change of dimension by changing the solvent quality. Figure 27A presents a ^{1}H-NMR spectrum of a brush with PS core and PAA shell, obtained in CD_3OD, which is a good solvent for the PAA shell, but a poor solvent for the PS core. The signals of the PAA shell are readily observed indicating an extended conformation, while no signals of PS are observed, implying the complete collapse of the PS core. By adding increasing amounts of $CDCl_3$ to this solution, one can restore the PS signals in the spectrum. (Fig. 27b). The PS core is solvated by $CDCl_3$ and therefore remains in an extended conformation. However, the worm-like structure of the micelle is preserved because of the covalent attachment of the PS-*b*-PAA side chains to the backbone. This behavior is very different from that of conventional dynamic micelles from linear amphiphilic diblock copolymers [127] which are expected to de-aggregate or form inverse micelles by changing the solvent quality.

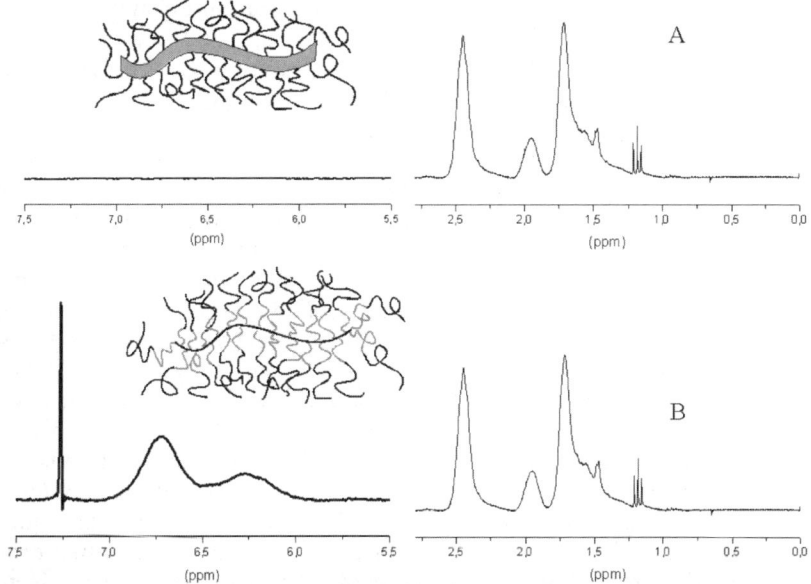

Fig. 27 ^{1}H-NMR spectra of [S$_{23}$-b-(AA)$_{186}$Br]$_{310}$, (**A**) in CD$_3$OD, and (**B**) in CD$_3$OD/CDCl$_3$ (volume ratio=1/1). Reprinted with permission from ref [99]. Copyright (2003) American Chemical Society

4.3
Graft Copolymers

The structure of graft copolymers is determined by three parameters: (i) the length of the backbone, (ii) the length and (iii) the *average* spacing of the side chains. Similar to the core-shell brushes, "grafting onto", "grafting from", or "grafting through", i.e., copolymerization of macromonomers with low-molecular-weight monomers can be used to synthesize graft copolymers. Living polymerizations techniques allow for controlling all three parameters. However, no convenient technique is available to control the spacing *distribution* which typically is quite broad.

Amphiphilic graft copolymers can be have either a hydrophobic backbone and hydrophilic/ionic side-chains or, inversely, a hydrophilic/ionic backbone and hydrophobic side-chains. The latter structures can form reversible hydrogels above a critical concentration and may find use as associative thickeners. Poly(acrylic acid)-*graft*-polystyrene (PAA-*g*-PS) graft copolymers were synthesized via condensation of NH$_2$-termitated PS with PAA, and the micellar behavior of these copolymers in aqueous media were investigated by Webber's group [128]. Light scattering and TEM investigations revealed that the micelle size relatively weakly depends on the grafting density, but strongly depends on the ionic strength of aqueous media. This type of polymers is reviewed in detail in the preceeding volume [129].

Not very much has been reported about the solution behavior of amphiphilic graft copolymers with hydrophobic backbone and ionic side-chains. Selb and Gallot [130, 131] synthesized polystyrene-*graft*-poly(4-*N*-ethyl vinylpyridinium bromide), PS-*g*-P4VPQ, via a coupling reaction of partially chloromethylated polystyrene with poly(4-vinylpyridine) anions, followed by quaternization of the pyridine units with ethyl bromide. The side-chains are strong cationic polyelectrolytes. The authors studied the viscosity in water/methanol mixed solutions and compared the data to those obtained with the corresponding block copolymers. With increasing water content the authors found micelle formation of the block copolymers whereas only unimers were observed for the graft copolymers. The authors assumed that the side-chains efficiently screen the collapsed backbone and thus keep it in solution in unimeric form. However, since polystyrene forms glassy, "frozen" micellar cores it is not clear whether these unimers are the thermodynamically stable forms in water. Non-quaternized PS-*g*-P2VP graft copolymers were also synthesized in the same way, and their aqueous solutions were characterized by Gauthier's group [132]. They found that these copolymers have a highly compact structure and expand much more in aqueous solution than the linear homologous polymers of poly(2-vinylpyridine) when protonated with HCl. This phenomenon is attributed to the higher charge density of the branches.

Poly(ethylene oxide)-*block*-poly(propylene oxide)-*block*-poly(ethylene oxide)-*g*-poly(acrylic acid) (PEO-*b*-PPO-*b*-PEO-*g*-PAA, Pluronic-PAA) graft copolymers were synthesized by free radical grafting copolymerization of acrylic acid monomers onto PEO-*b*-PPO-*b*-PEO (Pluronic F127) and the aqueous solution properties were characterized by Bromberg [133, 134]. Chiu et al. [135] reported on the micellization of (non-ionic) poly(stearyl methacrylate)-*graft*-poly(ethylene glycol) graft copolymers.

Recently, Müller et al. studied block and graft copolymers poly(n-butyl acrylate)-*block/graft*-poly(acrylic acid), PnBA-*b/g*-PAA [136]. The non-polar block/backbone has a low glass transition temperature, thus dynamic micelles were expected; the ionic block/side-chains are weak anionic polyelectrolytes, thus a strong dependence of micellization on pH could be expected. The graft copolymers were synthesized by ATRP copolymerization of poly(*tert*-butyl acrylate) macromonomers with *n*-butyl acrylate, followed by hydrolysis of the *tert*-butyl acrylate side-chains to PAA [137]. The length of the PAA side chains was varied from 20 to 85 monomer units and their number from 1.5 to 10, whereas the length of the backbone was kept at ca. 130 units.

Tensiometry, fluorometry, dynamic light scattering (DLS) and small-angle neutron scattering (SANS) studies revealed a strong dependence of the micellar size and the number of molecules per micelle on concentration, pH, ionic strength, and topology. Typically, micelles are only formed for pH$\leq$6, as can be seen from the hydrodynamic radii obtained by DLS (Fig. 28). This is easily explained by the fact that PAA is virtually non-ionized for pH $\leq$ 5, decreasing the solubility of the side-chains. This onset is at a somewhat higher pH for block copolymers. As expected, the micelle sizes increase with ionic strength due to screening of charges. Comparison of the number of

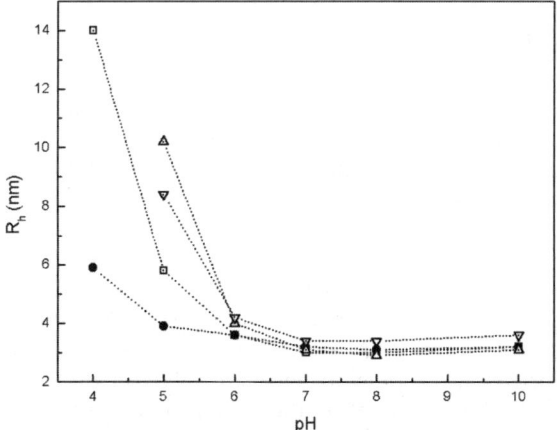

Fig. 28 pH dependence of the aggregation behavior of PnBA-g-PAA graft copolymers in 0.1 M NaCl aqueous solution at room temperature; (●) nBA$_{134}$-g-(AA$_{85}$)$_2$; (□) nBA$_{105}$-g-(AA$_{20}$)$_4$; (▽) nBA$_{170}$-g-(AA$_{20}$)$_3$, (△) nBA$_{150}$-g-(AA$_{20}$)$_{1.5}$

molecules per micelle, N_{agg}, (obtained by SANS at pH 5 in water with 0.1 M NaCl) for different fractions of acrylic acid unambiguously demonstrates that this number is significantly higher for block copolymers than for graft copolymers (Fig. 29). The explanation is based on the assumption that the collapsed hydrophobic core of unimers is better stabilized by many short arms than by one long block. In addition, the size of the core is limited by the size of the segments between two side-chains which is smaller than that of one block.

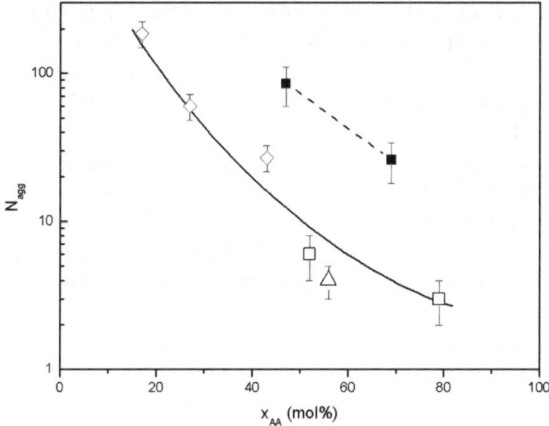

Fig. 29 Dependence of the degree of micellisation on the molar fraction of acrylic acid and topology: (◇) AA$_{20}$ side chains, (□) AA$_{37}$ side chains, (△) AA$_{85}$ side chains, (■) block copolymers

5
Conclusions and Outlook

The properties of polyelectrolyte block copolymer micelles have been investigated in detail in recent years. Due to the synthesis of well-defined polyelectrolyte block copolymers and the use of a variety of different experimental techniques the knowledge about polyelectrolyte block copolymer micelles and vesicles has well advanced.

Of central interest was the question of how the micellar core and polyelectrolyte shell structure depend on parameters such as block lengths, ionic strength and pH. It is observed that the grafting distance b has a characteristic dependence on the block length and decreases with addition of salt as it screens the electrostatic repulsion between polyelectrolyte chains. Depending on the added salt concentration the polyelectrolyte shell shows either the characteristic *"osmotic brush"* or *"salted brush"* behavior. In the *"osmotic brush"* regime the polyelectrolyte chains are strongly stretched and the micellar properties do not depend on the added salt concentration. In the *"salted brush"* regime where the added salt concentrations exceeds the internal concentration of counterions, the thickness of the polyelectrolyte shell decreases. The relation between added and internal salt concentration follows the classical *Donnan equilibrium*.

Due to the hydrophobicity of the polyelectrolyte backbone and the strong curvature of the polyelectrolyte layer the shell shows a tendency to phase separate into a dense, nearly homogeneous interior part and a dilute, diffuse outer part. The hydrophobic interactions together with fluctuations in the segment density lead to short-range attractive interactions. This can lead to the formation of micellar *networks* were micelles are either connected by *filaments* or are directly connected via overlapping layers forming doublets, higher multiplets or strings. at high micelle and salt concentration the networks grow in size and eventually (macro)phase separate into an isotropic micellar gel and a dilute micellar phase. Not only spherical micelles and networks, but also *vesicles* are formed by polyelectrolyte block copolymers. Vesicles can be dissolved by a decrease of the pH which is of relevance for the controlled release of drugs.

The range where repulsive interactions are dominant is the regime of *electrosteric stabilization*. The corresponding interaction potential can reasonably been assumed to be of a simple Debye-Hückel form. Measurements of the shear modulus of micellar gels as a function of micellar distance can well be described by this potential. The repulsion is one order of magnitude larger compared to purely steric interactions.

The results presented in literature so far on micelles of *ABC triblock copolymers* (or other superstructures of ABC triblock copolymers in solution) can only considered to be preliminary ones. In order to get a better understanding (and possibly control!) on these superstructures more systematic investigations are required. Due to the difficult synthesis and the very complicated solution behavior such tasks will not be easy, in general. An important question for future research is the establishment of scaling laws for the micellar diameter as a function of the degree of polymerization of the core, shell and corona blocks, respectively. Other questions deal with the interplay of the cationic and anionic blocks in dependence of their respective charge density which makes studies on block copolymer systems containing blocks with different acidity/basicity essential. Another important issue in this field is the unambiguous characterization of the solution morphologies. So far, mostly aqueous solutions have been investigated. However, also superstructures in organic solvents are interesting to study and so improvements in cryo-TEM will be helpful for their characterization. Recently cryo-TEM investigations of phospholipids in oil were published by Talmon's group [138].

Finally, it is shown that non-linear amphiphilic structures show different aggregation behavior as compared to block copolymers. Graft copolymers with non-polar backbone polyelectrolyte side chains have a smaller tendency to form micelles than their block copolymer analogs which is attributed to the more facile stabilization of unimers by the sidechains. In contrast, unimolecular micelles are the only possibility for core-shell nanoparticles. *Janus micelles*, on the other hand, form unique non-centrosymmetrical micelles that have a strong tendency to form centrosymmetrical supermicelles.

References

1. Selb J, Gallot Y (1985) In: Goodman I (ed) Developments in Block Copolymers, 2nd edn. Elsevier, Amsterdam
2. Selb J, Gallot Y (1980) Makromol Chem 181:809
3. Tuzar Z (1996) In: Webber SE, Munk P, Tuzar Z (eds) Solvents and Self-Organization of Polymers. NATO ASI Series E 327. Kluwer, Dordrecht
4. Valint L, Bock J (1988) Macromolecules 21:175
5. Astafieva I, Zhong XF, Eisenberg A (1993) Macromolecules 26:7339
6. Astafieva I, Khougaz K, Zhong XF, Eisenberg A (1995) Macromolecules 28:7127
7. Zhang LF, Eisenberg A (1995) Science 268:1728
8. Zhang LF, Barlow RJ, Eisenberg A (1995) Macromolecules 28:6055
9. Khougaz K, Astafieva I, Zhong XF, Eisenberg A (1995) Macromolecules 28, 7135
10. Gao Z, Varshney SK, Wong S, Eisenberg A (1994) Macromolecules 27:7923
11. Moffitt M, Khougaz K, Eisenberg A (1996) Acc Chem Res 29:95
12. Guenoun P, Davis HT, Tirrell M, Mays JW (1996) Macromolecules 29:3965
13. Guenoun P, Muller F, Delsanti M, Auvray L, Chen YJ, Mays JW, Tirrell M (1998) Phys Rev Lett 81:3872
14. Tuzar Z, Kratochvil P (1993) In: Matijevic E (ed) Surface and Colloid Science. Plenum, New York
15. Gast A, NATO ASI Ser. E 1998, 303, 311
16. Chu B (1995) Langmuir 11:414
17. Alexandridis P (1996) Curr Opin Colloid Interface Sci 1:490
18. Moffitt M, Khougaz K, Eisenberg A (1996) Acc Chem Res 29:95
19. Plantenberg T (2001) Ph.D. thesis, University of Hamburg
20. Förster S, Zisenis M, Wenz E, Antonietti M (1996) J Chem Phys 104:9956
21. Förster S, Plantenberg T (2002) Angew Chem Int Ed 41:688
22. Qin A, Tian M, Ramireddy C, Webber SE, Munk P, Tuzar Z (1994) Macromolecules 27:120
23. Eckert, AR, Webber SE (1996) Macromolecules 29:560

24. Voulgaris D, Tsitsilianis C, Grayer V, Esselink E, Hadziioannou G (1990) Polymer 40:5879
25. Voulgaris D, Tsitsilianis C (2000) Macromol Chem Phys in press
26. Marko JF, Rabin Y (1992) Macromolecules 25:1503
27. Dan N, Tirrell M (1993) Macromolecules 26:4310
28. Shusharina NP, Nyrkova IA, Khokhlov AR (1996) Macromolecules 29:3167
29. Förster S, Plantenberg T, Lindner P, submitted
30. Förster S, Schmidt M (1995) Adv Polym Sci 120:50
31. Barrat JL, Joanny JF (1996) Adv Chem Phys 94:1
32. Miklavi SJ, Marcelja S (1988) J Phys Chem 92:6718
33. Misra S, Varanasi S, Varanasi PP (1989) Macromolecules 22:4173
34. Zhulina EB, Borisov OV, Birshtein TM (1992) J Phys II (Paris) 2:63
35. Misra S, Mattice WL, Napper DH (1994) Macromolecules 27:7090
36. Zhulina EB (1996) In: Webber SE, Munk P, Tuzar Z (eds) In Solvents and Self-Organization of Polymers. NATO ASI Series E 327; Kluwer, Dordrecht
37. Pincus P (1991) Macromolecules 24:2912
38. Argillier JF, Tirrell M (1992) Theor Chim Acta 82, 343
39. Guenoun P, Schlachli A, Sentenac D, Mays JW, Benattar JJ (1995) Phys Rev Lett 74:3628
40. Hariharan R, Biver C, Mays JW, Russell WB (1998) Macromolecules 31:7506
41. Hariharan R, Biver C, Russell WB (1998) Macromolecules 31:7514
42. Tauer K, Müller H, Rosengarten L, Riedelsberger K (1999) Colloids Surfaces A: Physicochem. Eng. Aspects 153:75
43. Ahrens H, Förster S, Helm CA (1998) Macromolecules 30:8447
44. Ahrens H, Förster S, Helm CA (1998) Phys Rev Lett 81:4172
45. Förster S, Hermsdorf N, Leube W, Schnablegger H, Regenbrecht M, Akari S, Lindner P, Böttcher C (1999) J Phys Chem B 103:6652
46. Regenbrecht M, Akari S, Förster S, Möhwald H (1999) J Phys Chem B 103:6668
47. Regenbrecht M, Akari S, Förster S, Möhwald H (1999) Surface & Interface Analysis 27:418
48. Regenbrecht M, Akari S, Förster S, Netz RR, Möhwald H (1999) Nanotechnology 10:434
49. Förster S, Hermsdorf N, Böttcher C, Lindner P (2002) Macromolecules 35:4096
50. Tuzar Z, Prochazka K, Zuskova I, Munk P (1993) Polym Prepr 31:1038
51. Prochazka K, Kiserow D, Ramireddy C, Webber SE, Munk P, Tuzar Z (1992) Makromol Chem, Macromol Symp. 58:201
52. Stepanek M, Podhajecka K, Prochazka K, Teng Y, Webber SE (1999) Langmuir 15:4185
53. Stepanek M, Prochazka K, Brown W (2000) Langmuir 16:2502
54. Tsitsilianis C, Voulgaris D, Stepanek M, Podhajecka K, Prochazka K, Tuzar Z, Brown W (2000) Langmuir 16:6868
55. Kriz J, Basar B, Pospisil H, Plestil J, Tuzar Z, Kiselev MA (1996) Macromolecules 29:7853
56. Förster S, Plantenberg T, Hermsdorf N, Böttcher C, Lindner P, submitted
57. Liu et al (1987) J Cell Biol 104:527
58. Spatz JP, Roescher A., Sheiko S, Krausch G, Möller M (1995) Adv Mater 7:731
59. Li Z, Zhao W, Liu Y, Rafailovich MH, Sokolov J, Khougaz K, Eisenberg A, Lennox RB, Krausch G (1996) J Am Chem Soc 118:10892
60. Meiners JC, Quintelritzi A, Mlynek J, Elbs H, Krausch G (1997) Macromolecules 30:4945
61. Buitenhuis J, Förster S (1997) J Chem Phys 107:262
62. Witten TA, Pincus PA (1986) Macromolecules 19:2509
63. Förster S, Plantenberg T, Lindner P, submitted
64. Krämer E, Förster S, Göltner C, Antonietti M (1998) Langmuir 14:2027
65. Förster S, Borchert U, Lipprandt U, Lindner P, Funari S, submitted
66. Patrickios CS, Forder C, Armes SP, Billingham NC (1997) J Polym Sci, Part A: Polym Chem 35:1181

67. Kriz J, Masar B, Plestil J, Tuzar Z, Pospisil H, Doskocilova D (1998) Macromolecules 31:41
68. Yu G, Eisenberg A (1998) Macromolecules 31:5546
69. Ishizone T, Sugiyama K, Sakano Y, Mori H, Hirao A, Nakahama S (1999) Polymer J (Tokyo) 31:983
70. Patrickios CS, Hertler WR, Abbott NL, Hatton TA (1994) Macromolecules 27:930
71. Chen W-Y, Alexandridis P, Su C-K, Patrickios CS, Hertler WR, Hatton TA (1995) Macromolecules 28:8604
72. Patrickios CS, Lowe AB, Armes SP, Billingham NC (1998) J Polym Sci, Part A: Polym Chem 36:617
73. Giebeler E, Stadler R (1997) Macromol Chem Phys 198:3815
74. Bieringer R, Abetz V, Müller AHE (2001) Eur Phys J E: Soft Matter 5:5
75. Barthet C, Hickey AJ, Cairns DB, Armes SP (1999) Adv Mater (Weinheim, Ger.) 11:408
76. Alfrey T, Jr.; Fuoss, Raymond M.; Morawetz, Herbert; Pinner, Harry (1952) J Am Chem Soc 74:438
77. Alfrey T Jr, Pinner SH (1957) J Polymer Sci 23
78. Matyjaszewski K, Beers KL, Muhlebach A, Coca S, Zhang X, Gaynor SG (1998) Polym Mater Sci Eng 79:429
79. Ramireddy C, Tuzar Z, Prochazka K, Webber SE, Munk P (1992) Macromolecules 25:2541
80. Alfrey T Jr, Morawetz, Herbert (1952) J Am Chem Soc 74:436
81. Katchalsky AM, Israel R (1954) J Polymer Sci 13:57
82. Kamachi M, Kurihara M, Stille JK (1972) Macromolecules 5:161
83. Briggs NP, Budd PM, Price C (1992) Eur Polym J 28:739
84. Smid J, Fish D (1988) In: Mark HF (ed) Encyclopedia of Polymer Science and Engineering, vol 11. Wiley, New York, p 720
85. Tsuchida E, Osada Y, Ohno H (1980) J Macromol Sci-Phys B17:683
86. Bekturov EA, Bimendina LA (1981) Adv Polym Sci 41:99
87. Philipp B, Dawydoff W, Linow K-J (1982) Z Chem 22:1
88. Tsuchida E (1994) J Macromol Sci-Pure Appl Chem A31:1
89. Katchalsky A, Spitnik P (1947) J Polym Sci 2:432
90. Katchalsky A, Spitnik P (1947) J Polym Sci 2:487
91. Mazur J, Silberberg A, Katchalsky A (1959) J Polym Sci 35:43
92. Gohy J-F, Willet N, Varshney SK, Zhang J-X, Jérôme R (2002) e-Polymers no. 35
93. Gohy J-F, Willet N, Varshney S, Zhang J-X, Jérôme R (2001) Angewandte Chemie, International Edition 40:3214
94. Stewart S, Liu G (1999) Chem Materials 11:1048
95. Erhardt R, Böker A, Zettl H, Kaya H, Pyckhout-Hintzen W, Krausch G, Abetz V, Müller AHE (2001) Macromolecules 34:1069
96. Saito R, Fujita A, Ichimura A, Ishizu K (2000) J Polym Sci, Polym Chem Ed 38:2091
97. Xu H, Erhardt R, Abetz V, Müller AHE, Goedel EA (2001) Langmuir 17:6787
98. Erhardt R (2001) Janus-Micellen: Amphiphile oberflächenkompartimentierte Polymermicellen mit vernetztem Kern. Shaker, Aachen
99. Erhardt R, Zhang M, Böker A, Zettl H, Abetz C, Frederik P, Krausch G, Abetz V, Müller AHE (2003) J Am Chem Soc 125:3260
100. Cates ME, Candau SJ (1990) J Phys: Condens Matter 2:6869
101. Munk P, Ramireddy C, Tian M, Webber SE, Prochazka K, Tuzar Z (1992) Makromol Chem, Macromol Symp 58:195
102. Moffitt M, Khougaz K, Eisenberg A (1996) Accounts of Chemical Research 29:95
103. Förster S, Berton B, Hentze HP, Krämer E, Antonietti M, Lindner P (2001) Macromolecules 4610
104. Wang X-S, Winnik MA, Manners I (2002) Macromolecular Rapid Communications 23:210
105. Tsukahara Y, Mizuno K, Segawa A, Yamashita Y (1989) Macromolecules 22:1546
106. Tsukahara Y, Tsutsumi K, Yamashita Y, Shimada S (1990) Macromolecules 23:5201

107. Wintermantel M, Schmidt M, Tsukahara Y, Kajiwara K, Kohjiya S (1994) Macromol Rapid Commun. 15:279
108. Wintermantel M, Gerle M, Fischer K, Schmidt M, Wataoka I, Urakawa H, Kajiwara K, Tsukahara Y (1996) Macromolecules 29:978
109. Sheiko SS, Gerle M, Fischer K, Schmidt M, Möller M (1997) Langmuir 13:5368
110. Dziezok P, Sheiko SS, Fischer K, Schmidt M, Möller M (1998) Angew Chem, Int Ed Engl 36:2812
111. Kawaguchi S, Akaike K, Zhang Z-M, Matsumoto H, Ito K (1998) Polymer Journal (Tokyo) 30:1004
112. Tsukahara Y, Inoue J, Ohta Y, Kohjiya S, Okamoto Y (1994) Polym J 26:1013
113. Feast WJ, Gibson VC, Johnson AF, Khosravi E, Mohsin MA (1994) Polymer 35:3542
114. Heroguez V, Breunig S, Gnanou Y, Fontanille M (1996) Macromolecules 29:4459
115. Heroguez V, Gnanou Y, Fontanille M (1997) Macromolecules 307:4791
116. Schappacher M, Billaud C, Paulo C, Deffieux A (1999) Macromolecular Chemistry and Physics 200:2377
117. Deffieux A, Schappacher M (1999) Macromolecules 32:1797
118. Ryu SW, Hirao A (2000) Macromolecules 33:4765
119. Wang J-S, Matyjaszewski K (1995) J Am Chem Soc 117:5614
120. Beers KL, Gaynor SG, Matyjaszewski K, Sheiko SS, Möller M (1998) Macromolecules 31:9413
121. Cheng G, Böker A, Zhang M, Krausch G, Müller AHE (2001) Macromolecules 34:6883
122. Börner HG, Beers K, Matyjaszewski K, Sheiko SS, Möller M (2001) Macromolecules 34:4375
123. Djalali R, Hugenberg N, Fischer K, Schmidt M (1999) Macromol Rapid Commun. 20:444
124. Tsubaki K, Ishizu K (2001) Polymer 42:8387
125. Zhang M, Breiner T, Mori H, Müller AHE (2002) submitted
126. Djalali R, Li SY, Schmidt M (2002) Macromolecules 35:4282
127. Zhang LF, Shen HW, Eisenberg A (1997) Macromolecules 30:1001
128. Ma Y, Cao T, Webber SE (1998) Macromolecules 31:1773
129. Bohrisch J, Eisenbach CD, Jaeger W, Mori H, Müller AHE, Rehahn M, Schaller C, Traser S, Wittmeyer P (2003) Adv Polym Sci 165 (in press)
130. Selb J, Gallot Y (1979) Polymer 20:1273
131. Selb J, Gallot Y (1981) Makromol Chem 182:1775
132. Kee RA, Gauthier M (2002) Macromolecules 35:6526
133. Bromberg LJ (1998) Phys Chem B 102:1956
134. Bromberg LJ (1998) Phys Chem B 102:10736
135. Chiu H-C, Chern C-S, Lee C-K, Chang H-F (1998) Polymer 39:1609
136. Cai Y, Hartenstein M, Gradzielski M, Zhang M, Mori H, Pergushov DV, Lindner P, Zipfel J, Borisor O, Müller AHE (2002) submitted
137. Hartenstein M, Cai Y, Müller AHE (2002) submitted
138. Danino D, Gupta R, Satyavolu J, Talmon Y (2002) Journal of Colloid Interface Science 249:180

Received: November 2002

Author Index Volumes 101–166

Author Index Volumes 1-100 see Volume 100

de, Abajo, J. and *de la Campa, J. G.*: Processable Aromatic Polyimides.Vol. 140, pp. 23-60.
Abetz, V. see Förster, S.: Vol. 166, pp. 173-210.
Adolf, D. B. see Ediger, M. D.: Vol. 116, pp. 73-110.
Aharoni, S. M. and *Edwards, S. F.*: Rigid Polymer Networks.Vol. 118, pp. 1-231.
Albertsson, A.-C., Varma, I. K.: Aliphatic Polyesters: Synthesis, Properties and Applications. Vol. 157, pp. 99-138.
Albertsson, A.-C. see Edlund, U.: Vol. 157, pp. 53-98.
Albertsson, A.-C. see Söderqvist Lindblad, M.: Vol. 157, pp. 139-161.
Albertsson, A.-C. see Stridsberg, K. M.: Vol. 157, pp. 27-51.
Améduri, B., Boutevin, B. and *Gramain, P.*: Synthesis of Block Copolymers by Radical Polymerization and Telomerization. Vol. 127, pp. 87-142.
Améduri, B. and *Boutevin, B.*: Synthesis and Properties of Fluorinated Telechelic Monodispersed Compounds. Vol. 102, pp. 133-170.
Amselem, S. see Domb, A. J.: Vol. 107, pp. 93-142.
Andrady, A. L.: Wavelenght Sensitivity in Polymer Photodegradation. Vol. 128, pp. 47-94.
Andreis, M. and *Koenig, J. L.*: Application of Nitrogen-15 NMR to Polymers.Vol. 124, pp. 191-238.
Angiolini, L. see Carlini, C.: Vol. 123, pp. 127-214.
Anjum, N. see Gupta, B.: Vol. 162, pp. 37-63.
Anseth, K. S., Newman, S. M. and *Bowman, C. N.*: Polymeric Dental Composites: Properties and Reaction Behavior of Multimethacrylate Dental Restorations. Vol. 122, pp. 177-218.
Antonietti, M. see Cölfen, H.: Vol. 150, pp. 67-187.
Armitage, B. A. see O'Brien, D. F.: Vol. 126, pp. 53-58.
Arndt, M. see Kaminski, W.: Vol. 127, pp. 143-187.
Arnold Jr., F. E. and *Arnold, F. E.*: Rigid-Rod Polymers and Molecular Composites. Vol. 117, pp. 257-296.
Arora, M. see Kumar, M. N. V. R.: Vol. 160, pp. 45-118.
Arshady, R.: Polymer Synthesis via Activated Esters:A New Dimension of Creativity in Macromolecular Chemistry. Vol. 111, pp. 1-42.

Bahar, I., Erman, B. and *Monnerie, L.*: Effect of Molecular Structure on Local Chain Dynamics: Analytical Approaches and Computational Methods. Vol. 116, pp. 145-206.
Ballauff, M. see Dingenouts, N.: Vol. 144, pp. 1-48.
Ballauff, M. see Holm, C.: Vol. 166, pp. 1-27.
Ballauff, M. see Rühe, J.: Vol. 165, pp. 79-150.
Baltá-Calleja, F. J., González Arche, A., Ezquerra, T. A., Santa Cruz, C., Batallón, F., Frick, B. and *López Cabarcos, E.*: Structure and Properties of Ferroelectric Copolymers of Poly(vinylidene) Fluoride. Vol. 108, pp. 1-48.
Barnes, M. D. see Otaigbe, J.U.: Vol. 154, pp. 1-86.
Barshtein, G. R. and *Sabsai, O. Y.*: Compositions with Mineralorganic Fillers.Vol. 101, pp. 1-28.
Baschnagel, J., Binder, K., Doruker, P., Gusev, A. A., Hahn, O., Kremer, K., Mattice, W. L., Müller-Plathe, F., Murat, M., Paul, W., Santos, S., Sutter, U. W., Tries, V.: Bridging the Gap

Between Atomistic and Coarse-Grained Models of Polymers: Status and Perspectives. Vol. 152, pp. 41-156.
Batallán, F. see Baltá-Calleja, F. J.: Vol. 108, pp. 1-48.
Batog, A. E., Pet'ko, I.P., Penczek, P.: Aliphatic-Cycloaliphatic Epoxy Compounds and Polymers. Vol. 144, pp. 49-114.
Barton, J. see Hunkeler, D.: Vol. 112, pp. 115-134.
Bell, C. L. and *Peppas, N. A.:* Biomedical Membranes from Hydrogels and Interpolymer Complexes. Vol. 122, pp. 125-176.
Bellon-Maurel, A. see Calmon-Decriaud, A.: Vol. 135, pp. 207-226.
Bennett, D. E. see O'Brien, D. F.: Vol. 126, pp. 53-84.
Berry, G. C.: Static and Dynamic Light Scattering on Moderately Concentrated Solutions: Isotropic Solutions of Flexible and Rodlike Chains and Nematic Solutions of Rodlike Chains. Vol. 114, pp. 233-290.
Bershtein, V. A. and *Ryzhov, V. A.:* Far Infrared Spectroscopy of Polymers. Vol. 114, pp. 43-122.
Bhargava R., Wang S.-Q., Koenig J. L: FTIR Microspectroscopy of Polymeric Systems. Vol. 163, pp. 137-191.
Biesalski, M.: see Rühe, J.: Vol. 165, pp. 79-150.
Bigg, D. M.: Thermal Conductivity of Heterophase Polymer Compositions.Vol. 119, pp. 1-30.
Binder, K.: Phase Transitions in Polymer Blends and Block Copolymer Melts: Some Recent Developments. Vol. 112, pp. 115-134.
Binder, K.: Phase Transitions of Polymer Blends and Block Copolymer Melts in Thin Films. Vol. 138, pp. 1-90.
Binder, K. see Baschnagel, J.: Vol. 152, pp. 41-156.
Bird, R. B. see Curtiss, C. F.: Vol. 125, pp. 1-102.
Biswas, M. and *Mukherjee, A.:* Synthesis and Evaluation of Metal-Containing Polymers. Vol. 115, pp. 89-124.
Biswas, M. and *Sinha Ray, S.:* Recent Progress in Synthesis and Evaluation of Polymer-Montmorillonite Nanocomposites. Vol. 155, pp. 167-221.
Bogdal, D., Penczek, P., Pielichowski, J., Prociak, A.: Microwave Assisted Synthesis, Crosslinking, and Processing of Polymeric Materials. Vol. 163, pp. 193-263.
Bohrisch, J., Eisenbach, C.D., Jaeger, W., Mori H., Müller A.H.E., Rehahn, M., Schaller, C., Traser, S., Wittmeyer, P.: New Polyelectrolyte Architectures. Vol. 165, pp. 1-41.
Bolze, J. see Dingenouts, N.: Vol. 144, pp. 1-48.
Bosshard, C.: see Gubler, U.: Vol. 158, pp. 123-190.
Boutevin, B. and *Robin, J. J.:* Synthesis and Properties of Fluorinated Diols. Vol. 102. pp. 105-132.
Boutevin, B. see Amédouri, B.: Vol. 102, pp. 133-170.
Boutevin, B. see Améduri, B.: Vol. 127, pp. 87-142.
Bowman, C. N. see Anseth, K. S.: Vol. 122, pp. 177-218.
Boyd, R. H.: Prediction of Polymer Crystal Structures and Properties. Vol. 116, pp. 1-26.
Briber, R. M. see Hedrick, J. L.: Vol. 141, pp. 1-44.
Bronnikov, S. V., Vettegren, V. I. and *Frenkel, S. Y.:* Kinetics of Deformation and Relaxation in Highly Oriented Polymers. Vol. 125, pp. 103-146.
Brown, H. R. see Creton, C.: Vol. 156, pp. 53-135.
Bruza, K. J. see Kirchhoff, R. A.: Vol. 117, pp. 1-66.
Budkowski, A.: Interfacial Phenomena in Thin Polymer Films: Phase Coexistence and Segregation. Vol. 148, pp. 1-112.
Burban, J. H. see Cussler, E. L.: Vol. 110, pp. 67-80.
Burchard,W.: Solution Properties of Branched Macromolecules. Vol. 143, pp. 113-194.

Calmon-Decriaud, A., Bellon-Maurel, V., Silvestre, F.: Standard Methods for Testing the Aerobic Biodegradation of Polymeric Materials.Vol 135, pp. 207-226.
Cameron, N. R. and *Sherrington, D. C.:* High Internal Phase Emulsions (HIPEs)-Structure, Properties and Use in Polymer Preparation.Vol. 126, pp. 163-214.
de la Campa, J. G. see de Abajo, J.: Vol. 140, pp. 23-60.
Candau, F. see Hunkeler, D.: Vol. 112, pp. 115-134.

Canelas, D. A. and *DeSimone, J. M.*: Polymerizations in Liquid and Supercritical Carbon Dioxide. Vol. 133, pp. 103-140.
Canva, M., Stegeman, G. I.: Quadratic Parametric Interactions in Organic Waveguides. Vol. 158, pp. 87-121.
Capek, I.: Kinetics of the Free-Radical Emulsion Polymerization of Vinyl Chloride. Vol. 120, pp. 135-206.
Capek, I.: Radical Polymerization of Polyoxyethylene Macromonomers in Disperse Systems. Vol. 145, pp. 1-56.
Capek, I.: Radical Polymerization of Polyoxyethylene Macromonomers in Disperse Systems. Vol. 146, pp. 1-56.
Capek, I. and *Chern, C.-S.*: Radical Polymerization in Direct Mini-Emulsion Systems. Vol. 155, pp. 101-166.
Cappella, B. see Munz, M.: Vol. 164, pp. 87-210.
Carlesso, G. see Prokop, A.: Vol. 160, pp. 119-174.
Carlini, C. and *Angiolini, L.*: Polymers as Free Radical Photoinitiators. Vol. 123, pp. 127-214.
Carter, K. R. see Hedrick, J. L.: Vol. 141, pp. 1-44.
Casas-Vazquez, J. see Jou, D.: Vol. 120, pp. 207-266.
Chandrasekhar, V.: Polymer Solid Electrolytes: Synthesis and Structure. Vol 135, pp. 139-206.
Chang, J. Y. see Han, M. J.: Vol. 153, pp. 1-36.
Chang, T.: Recent Advances in Liquid Chromatography Analysis of Synthetic Polymers. Vol. 163, pp. 1-60.
Charleux, B., Faust R.: Synthesis of Branched Polymers by Cationic Polymerization. Vol. 142, pp. 1-70.
Chen, P. see Jaffe, M.: Vol. 117, pp. 297-328.
Chern, C.-S. see Capek, I.: Vol. 155, pp. 101-166.
Chevolot, Y. see Mathieu, H. J.: Vol. 162, pp. 1-35.
Choe, E.-W. see Jaffe, M.: Vol. 117, pp. 297-328.
Chow, T. S.: Glassy State Relaxation and Deformation in Polymers. Vol. 103, pp. 149-190.
Chung, S.-J. see Lin, T.-C.: Vol. 161, pp. 157-193
Chung, T.-S. see Jaffe, M.: Vol. 117, pp. 297-328.
Cölfen, H. and *Antonietti, M.*: Field-Flow Fractionation Techniques for Polymer and Colloid Analysis. Vol. 150, pp. 67-187.
Comanita, B. see Roovers, J.: Vol. 142, pp. 179-228.
Connell, J. W. see Hergenrother, P. M.: Vol. 117, pp. 67-110.
Creton, C., Kramer, E. J., Brown, H. R., Hui, C.-Y.: Adhesion and Fracture of Interfaces Between Immiscible Polymers: From the Molecular to the Continuum Scale. Vol. 156, pp. 53-135.
Criado-Sancho, M. see Jou, D.: Vol. 120, pp. 207-266.
Curro, J. G. see Schweizer, K. S.: Vol. 116, pp. 319-378.
Curtiss, C. F. and *Bird, R. B.*: Statistical Mechanics of Transport Phenomena: Polymeric Liquid Mixtures. Vol. 125, pp. 1-102.
Cussler, E. L., Wang, K. L. and *Burban, J. H.*: Hydrogels as Separation Agents. Vol. 110, pp. 67-80.

Dalton, L. Nonlinear Optical Polymeric Materials: From Chromophore Design to Commercial Applications. Vol. 158, pp. 1-86.
Dautzenberg, H. see Holm, C.: Vol. 166, pp.113-171.
Davidson, J. M. see Prokop, A.: Vol. 160, pp.119-174.
DeSimone, J. M. see Canelas D. A.: Vol. 133, pp. 103-140.
DiMari, S. see Prokop, A.: Vol. 136, pp. 1-52.
Dimonie, M. V. see Hunkeler, D.: Vol. 112, pp. 115-134.
Dingenouts, N., Bolze, J., Pötschke, D., Ballauf, M.: Analysis of Polymer Latexes by Small-Angle X-Ray Scattering. Vol. 144, pp. 1-48.
Dodd, L. R. and *Theodorou, D. N.*: Atomistic Monte Carlo Simulation and Continuum Mean Field Theory of the Structure and Equation of State Properties of Alkane and Polymer Melts. Vol. 116, pp. 249-282.
Doelker, E.: Cellulose Derivatives. Vol. 107, pp. 199-266.

Dolden, J. G.: Calculation of a Mesogenic Index with Emphasis Upon LC-Polyimides. Vol. 141, pp. 189 -245.
Domb, A. J., Amselem, S., Shah, J. and Maniar, M.: Polyanhydrides: Synthesis and Characterization. Vol. 107, pp. 93-142.
Domb, A. J. see Kumar, M. N. V. R.: Vol. 160, pp. 45118.
Doruker, P. see Baschnagel, J.: Vol. 152, pp. 41-156.
Dubois, P. see Mecerreyes, D.: Vol. 147, pp. 1-60.
Dubrovskii, S. A. see Kazanskii, K. S.: Vol. 104, pp. 97-134.
Dunkin, I. R. see Steinke, J.: Vol. 123, pp. 81-126.
Dunson, D. L. see McGrath, J. E.: Vol. 140, pp. 61-106.
Dziezok, P. see Rühe, J.: Vol. 165, pp. 79-150.

Eastmond, G. C.: Poly(ε-caprolactone) Blends. Vol. 149, pp. 59-223.
Economy, J. and Goranov, K.: Thermotropic Liquid Crystalline Polymers for High Performance Applications. Vol. 117, pp. 221-256.
Ediger, M. D. and Adolf, D. B.: Brownian Dynamics Simulations of Local Polymer Dynamics. Vol. 116, pp. 73-110.
Edlund, U. Albertsson, A.-C.: Degradable Polymer Microspheres for Controlled Drug Delivery. Vol. 157, pp. 53-98.
Edwards, S. F. see Aharoni, S. M.: Vol. 118, pp. 1-231.
Eisenbach, C. D. see Bohrisch, J.: Vol. 165, pp. 1-41.
Endo, T. see Yagci, Y.: Vol. 127, pp. 59-86.
Engelhardt, H. and Grosche, O.: Capillary Electrophoresis in Polymer Analysis. Vol.150, pp. 189-217.
Engelhardt, H. and Martin, H.: Characterization of Synthetic Polyelectrolytes by Capillary Electrophoretic Methods. Vol. 165, pp. 211-247.
Erman, B. see Bahar, I.: Vol. 116, pp. 145-206.
Eschner, M. see Spange, S.: Vol. 165, pp. 43-78.
Estel, K. see Spange, S.: Vol. 165, pp. 43-78.
Ewen, B, Richter, D.: Neutron Spin Echo Investigations on the Segmental Dynamics of Polymers in Melts, Networks and Solutions. Vol. 134, pp. 1-130.
Ezquerra, T. A. see Baltá-Calleja, F. J.: Vol. 108, pp. 1-48.

Faust, R. see Charleux, B: Vol. 142, pp. 1-70.
Fekete, E. see Pukánszky, B: Vol. 139, pp. 109-154.
Fendler, J. H.: Membrane-Mimetic Approach to Advanced Materials. Vol. 113, pp. 1-209.
Fetters, L. J. see Xu, Z.: Vol. 120, pp. 1-50.
Förster, S., Abetz, V., Müller, A. H. E.: Polyelectrolyte Block Copolymer Micelles. Vol. 166, pp. 173-210.
Förster, S. and Schmidt, M.: Polyelectrolytes in Solution. Vol. 120, pp. 51-134.
Freire, J. J.: Conformational Properties of Branched Polymers: Theory and Simulations. Vol. 143, pp. 35-112.
Frenkel, S. Y. see Bronnikov, S.V.: Vol. 125, pp. 103-146.
Frick, B. see Baltá-Calleja, F. J.: Vol. 108, pp. 1-48.
Fridman, M. L.: see Terent'eva, J. P.: Vol. 101, pp. 29-64.
Fukui, K. see Otaigbe, J. U.: Vol. 154, pp. 1-86.
Funke, W.: Microgels-Intramolecularly Crosslinked Macromolecules with a Globular Structure. Vol. 136, pp. 137-232.

Galina, H.: Mean-Field Kinetic Modeling of Polymerization: The Smoluchowski Coagulation Equation.Vol. 137, pp. 135-172.
Ganesh, K. see Kishore, K.: Vol. 121, pp. 81-122.
Gaw, K. O. and Kakimoto, M.: Polyimide-Epoxy Composites. Vol. 140, pp. 107-136.
Geckeler, K. E. see Rivas, B.: Vol. 102, pp. 171-188.
Geckeler, K. E.: Soluble Polymer Supports for Liquid-Phase Synthesis. Vol. 121, pp. 31-80.
Gehrke, S. H.: Synthesis, Equilibrium Swelling, Kinetics Permeability and Applications of Environmentally Responsive Gels. Vol. 110, pp. 81-144.

de Gennes, P.-G.: Flexible Polymers in Nanopores. Vol. 138, pp. 91-106.
Geuss, M. see Munz, M.: Vol. 164, pp. 87-210
Giannelis, E. P., Krishnamoorti, R., Manias, E.: Polymer-Silicate Nanocomposites: Model Systems for Confined Polymers and Polymer Brushes. Vol. 138, pp. 107-148.
Godovsky, D. Y.: Device Applications of Polymer-Nanocomposites. Vol. 153, pp. 163-205.
Godovsky, D. Y.: Electron Behavior and Magnetic Properties Polymer-Nanocomposites. Vol. 119, pp. 79-122.
González Arche, A. see Baltá-Calleja, F. J.: Vol. 108, pp. 1-48.
Goranov, K. see Economy, J.: Vol. 117, pp. 221-256.
Gramain, P. see Améduri, B.: Vol. 127, pp. 87-142.
Grest, G. S.: Normal and Shear Forces Between Polymer Brushes. Vol. 138, pp. 149-184.
Grigorescu, G, Kulicke, W.-M.: Prediction of Viscoelastic Properties and Shear Stability of Polymers in Solution. Vol. 152, p. 1-40.
Gröhn, F. see Rühe, J.: Vol. 165, pp. 79-150.
Grosberg, A. and Nechaev, S.: Polymer Topology. Vol. 106, pp. 1-30.
Grosche, O. see Engelhardt, H.: Vol. 150, pp. 189-217.
Grubbs, R., Risse, W. and Novac, B.: The Development of Well-defined Catalysts for Ring-Opening Olefin Metathesis. Vol. 102, pp. 47-72.
Gubler, U., Bosshard, C.: Molecular Design for Third-Order Nonlinear Optics. Vol. 158, pp. 123-190.
van Gunsteren, W. F. see Gusev, A. A.: Vol. 116, pp. 207-248.
Gupta, B., Anjum, N.: Plasma and Radiation-Induced Graft Modification of Polymers for Biomedical Applications. Vol. 162, pp. 37-63.
Gusev, A. A., Müller-Plathe, F., van Gunsteren, W. F. and Suter, U. W.: Dynamics of Small Molecules in Bulk Polymers. Vol. 116, pp. 207-248.
Gusev, A. A. see Baschnagel, J.: Vol. 152, pp. 41-156.
Guillot, J. see Hunkeler, D.: Vol. 112, pp. 115-134.
Guyot, A. and Tauer, K.: Reactive Surfactants in Emulsion Polymerization. Vol. 111, pp. 43-66.

Hadjichristidis, N., Pispas, S., Pitsikalis, M., Iatrou, H., Vlahos, C.: Asymmetric Star Polymers Synthesis and Properties. Vol. 142, pp. 71-128.
Hadjichristidis, N. see Xu, Z.: Vol. 120, pp. 1-50.
Hadjichristidis, N. see Pitsikalis, M.: Vol. 135, pp. 1-138.
Hahn, O. see Baschnagel, J.: Vol. 152, pp. 41-156.
Hakkarainen, M.: Aliphatic Polyesters: Abiotic and Biotic Degradation and Degradation Products. Vol. 157, pp. 1-26.
Hall, H. K. see Penelle, J.: Vol. 102, pp. 73-104.
Hamley, I.W.: Crystallization in Block Copolymers. Vol. 148, pp. 113-138.
Hammouda, B.: SANS from Homogeneous Polymer Mixtures: A Unified Overview. Vol. 106, pp. 87-134.
Han, M. J. and Chang, J. Y.: Polynucleotide Analogues. Vol. 153, pp. 1-36.
Harada, A.: Design and Construction of Supramolecular Architectures Consisting of Cyclodextrins and Polymers. Vol. 133, pp. 141-192.
Haralson, M. A. see Prokop, A.: Vol. 136, pp. 1-52.
Hassan, C. M. and Peppas, N. A.: Structure and Applications of Poly(vinyl alcohol) Hydrogels Produced by Conventional Crosslinking or by Freezing/Thawing Methods. Vol. 153, pp. 37-65.
Hawker, C. J.: Dentritic and Hyperbranched Macromolecules Precisely Controlled Macromolecular Architectures. Vol. 147, pp. 113-160.
Hawker, C. J. see Hedrick, J. L.: Vol. 141, pp. 1-44.
He, G. S. see Lin, T.-C.: Vol. 161, pp. 157-193.
Hedrick, J. L., Carter, K. R., Labadie, J. W., Miller, R. D., Volksen, W., Hawker, C. J., Yoon, D. Y., Russell, T. P., McGrath, J. E., Briber, R. M.: Nanoporous Polyimides. Vol. 141, pp. 1-44.
Hedrick, J. L., Labadie, J. W., Volksen, W. and Hilborn, J. G.: Nanoscopically Engineered Polyimides. Vol. 147, pp. 61-112.
Hedrick, J. L. see Hergenrother, P. M.: Vol. 117, pp. 67-110.

Hedrick, J. L. see Kiefer, J.: Vol. 147, pp. 161-247.
Hedrick, J. L. see McGrath, J. E.: Vol. 140, pp. 61-106.
Heinrich, G. and *Klüppel, M.*: Recent Advances in the Theory of Filler Networking in Elastomers. Vol. 160, pp. 1-44.
Heller, J.: Poly (Ortho Esters). Vol. 107, pp. 41-92.
Helm, C. A.: see Möhwald, H.: Vol. 165, pp. 151-175.
Hemielec, A. A. see Hunkeler, D.: Vol. 112, pp. 115-134.
Hergenrother, P. M., Connell, J. W., Labadie, J. W. and *Hedrick, J. L.*: Poly(arylene ether)s Containing Heterocyclic Units. Vol. 117, pp. 67-110.
Hernández-Barajas, J. see Wandrey, C.: Vol. 145, pp. 123-182.
Hervet, H. see Léger, L.: Vol. 138, pp. 185-226.
Hilborn, J. G. see Hedrick, J. L.: Vol. 147, pp. 61-112.
Hilborn, J. G. see Kiefer, J.: Vol. 147, pp. 161-247.
Hiramatsu, N. see Matsushige, M.: Vol. 125, pp. 147-186.
Hirasa, O. see Suzuki, M.: Vol. 110, pp. 241-262.
Hirotsu, S.: Coexistence of Phases and the Nature of First-Order Transition in Poly-N-isopropylacrylamide Gels. Vol. 110, pp. 1-26.
Höcker, H. see Klee, D.: Vol. 149, pp. 1-57.
Holm, C., Hofmann, T., Joanny, J. F., Kremer, K., Netz, R. R., Reineker, P., Seidel, C., Vilgis, T. A., Winkler, R. G.: Polyelectrolyte Theory. Vol. 166, pp. 67-111.
Holm, C., Rehahn, M., Oppermann, W., Ballauff, M.: Stiff-Chain Polyelectrolytes. Vol. 166, pp. 1-27.
Hornsby, P.: Rheology, Compoundind and Processing of Filled Thermoplastics. Vol. 139, pp. 155-216.
Houbenov, N. see Rühe, J.: Vol. 165, pp. 79-150.
Huber, K. see Volk, N.: Vol. 166, pp. 29-65.
Hugenberg, N. see Rühe, J.: Vol. 165, pp. 79-150.
Hui, C.-Y. see Creton, C.: Vol. 156, pp. 53-135.
Hult, A., Johansson, M., Malmström, E.: Hyperbranched Polymers.Vol. 143, pp. 1-34.
Hunkeler, D., Candau, F., Pichot, C., Hemielec, A. E., Xie, T. Y., Barton, J., Vaskova, V., Guillot, J., Dimonie, M. V., Reichert, K. H.: Heterophase Polymerization: A Physical and Kinetic Comparision and Categorization. Vol. 112, pp. 115-134.
Hunkeler, D. see Macko, T.: Vol. 163, pp. 61-136.
Hunkeler, D. see Prokop, A.: Vol. 136, pp. 1-52; 53-74.
Hunkeler, D see Wandrey, C.: Vol. 145, pp. 123-182.

Iatrou, H. see Hadjichristidis, N.: Vol. 142, pp. 71-128.
Ichikawa, T. see Yoshida, H.: Vol. 105, pp. 3-36.
Ihara, E. see Yasuda, H.: Vol. 133, pp. 53-102.
Ikada, Y. see Uyama,Y.: Vol. 137, pp. 1-40.
Ilavsky, M.: Effect on Phase Transition on Swelling and Mechanical Behavior of Synthetic Hydrogels. Vol. 109, pp. 173-206.
Imai, Y.: Rapid Synthesis of Polyimides from Nylon-Salt Monomers. Vol. 140, pp. 1-23.
Inomata, H. see Saito, S.: Vol. 106, pp. 207-232.
Inoue, S. see Sugimoto, H.: Vol. 146, pp. 39-120.
Irie, M.: Stimuli-Responsive Poly(N-isopropylacrylamide), Photo- and Chemical-Induced Phase Transitions. Vol. 110, pp. 49-66.
Ise, N. see Matsuoka, H.: Vol. 114, pp. 187-232.
Ito, K., Kawaguchi, S.: Poly(macronomers), Homo- and Copolymerization. Vol. 142, pp. 129-178.
Ivanov, A. E. see Zubov, V. P.: Vol. 104, pp. 135-176.

Jacob, S. and *Kennedy, J.*: Synthesis, Characterization and Properties of OCTA-ARM Polyisobutylene-Based Star Polymers. Vol. 146, pp. 1-38.
Jaeger, W. see Bohrisch, J.: Vol. 165, pp. 1-41.
Jaffe, M., Chen, P., Choe, E.-W., Chung, T.-S. and *Makhija, S.*: High Performance Polymer Blends. Vol. 117, pp. 297-328.

Jancar, J.: Structure-Property Relationships in Thermoplastic Matrices. Vol. 139, pp. 1-66.
Jen, A. K-Y. see Kajzar, F.: Vol. 161, pp. 1-85.
Jerome, R. see Mecerreyes, D.: Vol. 147, pp. 1-60.
Jiang, M., Li, M., Xiang, M. and *Zhou, H.*: Interpolymer Complexation and Miscibility and Enhancement by Hydrogen Bonding. Vol. 146, pp. 121-194.
Jin, J. see Shim, H.-K.: Vol. 158, pp. 191-241.
Jo, W. H. and *Yang, J. S.*: Molecular Simulation Approaches for Multiphase Polymer Systems. Vol. 156, pp. 1-52.
Joanny, J.-F. see Holm, C.: Vol. 166, pp. 67-111.
Joanny, J.-F. see Thünemann, A. F.: Vol. 166, pp. 113-171.
Johannsmann, D. see Rühe, J.: Vol. 165, pp. 79-150.
Johansson, M. see Hult, A.: Vol. 143, pp. 1-34.
Joos-Müller, B. see Funke, W.: Vol. 136, pp. 137-232.
Jou, D., Casas-Vazquez, J. and *Criado-Sancho, M.*: Thermodynamics of Polymer Solutions under Flow: Phase Separation and Polymer Degradation. Vol. 120, pp. 207-266.

Kaetsu, I.: Radiation Synthesis of Polymeric Materials for Biomedical and Biochemical Applications. Vol. 105, pp. 81-98.
Kaji, K. see Kanaya, T.: Vol. 154, pp. 87-141.
Kajzar, F., Lee, K.-S., Jen, A. K.-Y.: Polymeric Materials and their Orientation Techniques for Second-Order Nonlinear Optics. Vol. 161, pp. 1-85.
Kakimoto, M. see Gaw, K. O.: Vol. 140, pp. 107-136.
Kaminski, W. and *Arndt, M.*: Metallocenes for Polymer Catalysis. Vol. 127, pp. 143-187.
Kammer, H. W., Kressler, H. and *Kummerloewe, C.*: Phase Behavior of Polymer Blends - Effects of Thermodynamics and Rheology. Vol. 106, pp. 31-86.
Kanaya, T. and *Kaji, K.*: Dynamcis in the Glassy State and Near the Glass Transition of Amorphous Polymers as Studied by Neutron Scattering. Vol. 154, pp. 87-141.
Kandyrin, L. B. and *Kuleznev, V. N.*: The Dependence of Viscosity on the Composition of Concentrated Dispersions and the Free Volume Concept of Disperse Systems. Vol. 103, pp. 103-148.
Kaneko, M. see Ramaraj, R.: Vol. 123, pp. 215-242.
Kang, E. T., Neoh, K. G. and *Tan, K. L.*: X-Ray Photoelectron Spectroscopic Studies of Electroactive Polymers. Vol. 106, pp. 135-190.
Karlsson, S. see Söderqvist Lindblad, M.: Vol. 157, pp. 139-161.
Kato, K. see Uyama,Y.: Vol. 137, pp. 1-40.
Kawaguchi, S. see Ito, K.: Vol. 142, p 129-178.
Kazanskii, K. S. and *Dubrovskii, S. A.*: Chemistry and Physics of Agricultural Hydrogels. Vol. 104, pp. 97-134.
Kennedy, J. P. see Jacob, S.: Vol. 146, pp. 1-38.
Kennedy, J. P. see Majoros, I.: Vol. 112, pp. 1-113.
Khokhlov, A., Starodybtzev, S. and *Vasilevskaya, V.*: Conformational Transitions of Polymer Gels: Theory and Experiment. Vol. 109, pp. 121-172.
Kiefer, J., Hedrick J. L. and *Hiborn, J. G.*: Macroporous Thermosets by Chemically Induced Phase Separation. Vol. 147, pp. 161-247.
Kilian, H. G. and *Pieper, T.*: Packing of Chain Segments. A Method for Describing X-Ray Patterns of Crystalline, Liquid Crystalline and Non-Crystalline Polymers. Vol. 108, pp. 49-90.
Kim, J. see Quirk, R.P.: Vol. 153, pp. 67-162.
Kim, K.-S. see Lin, T.-C.: Vol. 161, pp. 157-193.
Kippelen, B. and *Peyghambarian, N.*: Photorefractive Polymers and their Applications. Vol. 161, pp. 87-156.
Kishore, K. and *Ganesh, K.*: Polymers Containing Disulfide, Tetrasulfide, Diselenide and Ditelluride Linkages in the Main Chain. Vol. 121, pp. 81-122.
Kitamaru, R.: Phase Structure of Polyethylene and Other Crystalline Polymers by Solid-State 13C/MNR. Vol. 137, pp 41-102.
Klee, D. and *Höcker, H.*: Polymers for Biomedical Applications: Improvement of the Interface Compatibility. Vol. 149, pp. 1-57.

Klier, J. see Scranton, A. B.: Vol. 122, pp. 1-54.
v. Klitzing, R. and *Tieke, B.:* Polyelectrolyte Membranes. Vol. 165, pp. 177-210.
Klüppel, M.: The Role of Disorder in Filler Reinforcement of Elastomers on Various Length Scales. Vol. 164, pp. 1-86
Klüppel, M. see Heinrich, G.: Vol. 160, pp 1-44.
Kobayashi, S., Shoda, S. and *Uyama, H.:* Enzymatic Polymerization and Oligomerization. Vol. 121, pp. 1-30.
Köhler, W. and *Schäfer, R.:* Polymer Analysis by Thermal-Diffusion Forced Rayleigh Scattering. Vol. 151, pp. 1-59.
Koenig, J. L. see Bhargava, R.: Vol. 163, pp. 137-191.
Koenig, J. L. see Andreis, M.: Vol. 124, pp. 191-238.
Koike, T.: Viscoelastic Behavior of Epoxy Resins Before Crosslinking. Vol. 148, pp. 139-188.
Kokufuta, E.: Novel Applications for Stimulus-Sensitive Polymer Gels in the Preparation of Functional Immobilized Biocatalysts. Vol. 110, pp. 157-178.
Konno, M. see Saito, S.: Vol. 109, pp. 207-232.
Konradi, R. see Rühe, J.: Vol. 165, pp. 79-150.
Kopecek, J. see Putnam, D.: Vol. 122, pp. 55-124.
Koßmehl, G. see Schopf, G.: Vol. 129, pp. 1-145.
Kozlov, E. see Prokop, A.: Vol. 160, pp. 119-174.
Kramer, E. J. see Creton, C.: Vol. 156, pp. 53-135.
Kremer, K. see Baschnagel, J.: Vol. 152, pp. 41-156.
Kremer, K. see Holm, C.: Vol. 166, pp. 67-111.
Kressler, J. see Kammer, H. W.: Vol. 106, pp. 31-86.
Kricheldorf, H. R.: Liquid-Cristalline Polyimides. Vol. 141, pp. 83-188.
Krishnamoorti, R. see Giannelis, E. P.: Vol. 138, pp. 107-148.
Kirchhoff, R. A. and *Bruza, K. J.:* Polymers from Benzocyclobutenes. Vol. 117, pp. 1-66.
Kuchanov, S. I.: Modern Aspects of Quantitative Theory of Free-Radical Copolymerization. Vol. 103, pp. 1-102.
Kuchanov, S. I.: Principles of Quantitive Description of Chemical Structure of Synthetic Polymers. Vol. 152, p. 157-202.
Kudaibergennow, S. E.: Recent Advances in Studying of Synthetic Polyampholytes in Solutions. Vol. 144, pp. 115-198.
Kuleznev, V. N. see Kandyrin, L. B.: Vol. 103, pp. 103-148.
Kulichkhin, S. G. see Malkin, A. Y.: Vol. 101, pp. 217-258.
Kulicke, W.-M. see Grigorescu, G.: Vol. 152, p. 1-40.
Kumar, M. N. V. R., Kumar, N., Domb, A. J. and *Arora, M.:* Pharmaceutical Polymeric Controlled Drug Delivery Systems. Vol. 160, pp. 45-118.
Kumar, N. see Kumar M. N. V. R.: Vol. 160, pp. 45-118.
Kummerloewe, C. see Kammer, H. W.: Vol. 106, pp. 31-86.
Kuznetsova, N. P. see Samsonov, G.V.: Vol. 104, pp. 1-50.

Labadie, J. W. see Hergenrother, P. M.: Vol. 117, pp. 67-110.
Labadie, J. W. see Hedrick, J. L.: Vol. 141, pp. 1-44.
Labadie, J. W. see Hedrick, J. L.: Vol. 147, pp. 61-112.
Lamparski, H. G. see O'Brien, D. F.: Vol. 126, pp. 53-84.
Laschewsky, A.: Molecular Concepts, Self-Organisation and Properties of Polysoaps. Vol. 124, pp. 1-86.
Laso, M. see Leontidis, E.: Vol. 116, pp. 283-318.
Lazár, M. and *Rychl, R.:* Oxidation of Hydrocarbon Polymers. Vol. 102, pp. 189-222.
Lechowicz, J. see Galina, H.: Vol. 137, pp. 135-172.
Léger, L., Raphaël, E., Hervet, H.: Surface-Anchored Polymer Chains: Their Role in Adhesion and Friction. Vol. 138, pp. 185-226.
Lenz, R. W.: Biodegradable Polymers. Vol. 107, pp. 1-40.
Leontidis, E., de Pablo, J. J., Laso, M. and *Suter, U. W.:* A Critical Evaluation of Novel Algorithms for the Off-Lattice Monte Carlo Simulation of Condensed Polymer Phases. Vol. 116, pp. 283-318.

Lee, B. see Quirk, R. P.: Vol. 153, pp. 67-162.
Lee, K.-S. see Kajzar, F.: Vol. 161, pp. 1-85.
Lee, Y. see Quirk, R. P: Vol. 153, pp. 67-162.
Leónard, D. see Mathieu, H. J.: Vol. 162, pp. 1-35.
Lesec, J. see Viovy, J.-L.: Vol. 114, pp. 1-42.
Li, M. see Jiang, M.: Vol. 146, pp. 121-194.
Liang, G. L. see Sumpter, B. G.: Vol. 116, pp. 27-72.
Lienert, K.-W.: Poly(ester-imide)s for Industrial Use. Vol. 141, pp. 45-82.
Lin, J. and *Sherrington, D. C.*: Recent Developments in the Synthesis, Thermostability and Liquid Crystal Properties of Aromatic Polyamides. Vol. 111, pp. 177-220.
Lin, T.-C., Chung, S.-J., Kim, K.-S., Wang, X., He, G. S., Swiatkiewicz, J., Pudavar, H. E. and *Prasad, P. N.*: Organics and Polymers with High Two-Photon Activities and their Applications. Vol. 161, pp. 157-193.
Liu, Y. see Söderqvist Lindblad, M.: Vol. 157, pp. 139161
López Cabarcos, E. see Baltá-Calleja, F. J.: Vol. 108, pp. 1-48.
Löwen, H. see Thünemann, A. F.: Vol. 166, pp. 113-171.

Macko, T. and *Hunkeler, D.*: Liquid Chromatography under Critical and Limiting Conditions: A Survey of Experimental Systems for Synthetic Polymers. Vol. 163, pp. 61-136.
Majoros, I., Nagy, A. and *Kennedy, J. P.*: Conventional and Living Carbocationic Polymerizations United. I.A Comprehensive Model and New Diagnostic Method to Probe the Mechanism of Homopolymerizations. Vol. 112, pp. 1-113.
Makhija, S. see Jaffe, M.: Vol. 117, pp. 297-328.
Malmström, E. see Hult, A.: Vol. 143, pp. 1-34.
Malkin, A. Y. and *Kulichkhin, S. G.*: Rheokinetics of Curing. Vol. 101, pp. 217-258.
Maniar, M. see Domb, A. J.: Vol. 107, pp. 93-142.
Manias, E. see Giannelis, E. P.: Vol. 138, pp. 107-148.
Martin, H. see Engelhardt, H.: Vol. 165, pp. 211-247.
Mashima, K., Nakayama, Y. and *Nakamura, A.*: Recent Trends in Polymerization of a-Olefins Catalyzed by Organometallic Complexes of Early Transition Metals.Vol. 133, pp. 1-52.
Mathew, D. see Reghunadhan Nair, C.P.: Vol. 155, pp. 1-99.
Mathieu, H. J., Chevolot, Y, Ruiz-Taylor, L. and *Leónard, D.*: Engineering and Characterization of Polymer Surfaces for Biomedical Applications. Vol. 162, pp. 1-35.
Matsumoto, A.: Free-Radical Crosslinking Polymerization and Copolymerization of Multivinyl Compounds. Vol. 123, pp. 41-80.
Matsumoto, A. see Otsu, T.: Vol. 136, pp. 75-138.
Matsuoka, H. and *Ise, N.*: Small-Angle and Ultra-Small Angle Scattering Study of the Ordered Structure in Polyelectrolyte Solutions and Colloidal Dispersions. Vol. 114, pp. 187-232.
Matsushige, K., Hiramatsu, N. and *Okabe, H.*: Ultrasonic Spectroscopy for Polymeric Materials. Vol. 125, pp. 147-186.
Mattice, W. L. see Rehahn, M.: Vol. 131/132, pp. 1-475.
Mattice, W. L. see Baschnagel, J.: Vol. 152, p. 41-156.
Mays, W. see Xu, Z.: Vol. 120, pp. 1-50.
Mays, J. W. see Pitsikalis, M.: Vol. 135, pp. 1-138.
McGrath, J. E. see Hedrick, J. L.: Vol. 141, pp. 1-44.
McGrath, J. E., Dunson, D. L., Hedrick, J. L.: Synthesis and Characterization of Segmented Polyimide-Polyorganosiloxane Copolymers. Vol. 140, pp. 61-106.
McLeish, T. C. B., Milner, S. T.: Entangled Dynamics and Melt Flow of Branched Polymers. Vol. 143, pp. 195-256.
Mecerreyes, D., Dubois, P. and *Jerome, R.*: Novel Macromolecular Architectures Based on Aliphatic Polyesters: Relevance of the Coordination-Insertion Ring-Opening Polymerization. Vol. 147, pp. 1-60.
Mecham, S. J. see McGrath, J. E.: Vol. 140, pp. 61-106.
Menzel, H. see Möhwald, H.: Vol. 165, pp. 151-175.
Meyer, T. see Spange, S.: Vol. 165, pp. 43-78.
Mikos, A. G. see Thomson, R. C.: Vol. 122, pp. 245-274.

Milner, S. T. see McLeish, T. C. B.: Vol. 143, pp. 195-256.
Mison, P. and *Sillion, B.*: Thermosetting Oligomers Containing Maleimides and Nadiimides End-Groups. Vol. 140, pp. 137-180.
Miyasaka, K.: PVA-Iodine Complexes: Formation, Structure and Properties. Vol. 108. pp. 91-130.
Miller, R. D. see Hedrick, J. L.: Vol. 141, pp. 1-44.
Minko, S. see Rühe, J.: Vol. 165, pp. 79-150.
Möhwald, H., Menzel, H., Helm, C. A., Stamm, M.: Lipid and Polyampholyte Monolayers to Study Polyelectrolyte Interactions and Structure at Interfaces. Vol. 165, pp. 151-175.
Monnerie, L. see Bahar, I.: Vol. 116, pp. 145-206.
Mori, H. see Bohrisch, J.: Vol. 165, pp. 1-41.
Morishima, Y.: Photoinduced Electron Transfer in Amphiphilic Polyelectrolyte Systems. Vol. 104, pp. 51-96.
Morton M. see Quirk, R. P: Vol. 153, pp. 67-162.
Motornov, M. see Rühe, J.: Vol. 165, pp. 79-150.
Mours, M. see Winter, H. H.: Vol. 134, pp. 165-234.
Müllen, K. see Scherf, U.: Vol. 123, pp. 1-40.
Müller, A.H.E. see Bohrisch, J.: Vol. 165, pp. 1-41.
Müller, A.H.E. see Förster, S.: Vol. 166, pp. 173-210.
Müller, M. see Thünemann, A. F.: Vol. 166, pp. 113-171.
Müller-Plathe, F. see Gusev, A. A.: Vol. 116, pp. 207-248.
Müller-Plathe, F. see Baschnagel, J.: Vol. 152, p. 41-156.
Mukerherjee, A. see Biswas, M.: Vol. 115, pp. 89-124.
Munz, M., Cappella, B., Sturm, H., Geuss, M., Schulz, E.: Materials Contrasts and Nanolithography Techniques in Scanning Force Microscopy (SFM) and their Application to Polymers and Polymer Composites. Vol. 164, pp. 87-210
Murat, M. see Baschnagel, J.: Vol. 152, p. 41-156.
Mylnikov, V.: Photoconducting Polymers. Vol. 115, pp. 1-88.

Nagy, A. see Majoros, I.: Vol. 112, pp. 1-11.
Nakamura, A. see Mashima, K.: Vol. 133, pp. 1-52.
Nakayama, Y. see Mashima, K.: Vol. 133, pp. 1-52.
Narasinham, B., Peppas, N. A.: The Physics of Polymer Dissolution: Modeling Approaches and Experimental Behavior. Vol. 128, pp. 157-208.
Nechaev, S. see Grosberg, A.: Vol. 106, pp. 1-30.
Neoh, K. G. see Kang, E. T.: Vol. 106, pp. 135-190.
Netz, R.R. see Holm, C.: Vol. 166, pp. 67-111.
Netz, R.R. see Rühe, J.: Vol. 165, pp. 79-150.
Newman, S. M. see Anseth, K. S.: Vol. 122, pp. 177-218.
Nijenhuis, K. te: Thermoreversible Networks. Vol. 130, pp. 1-252.
Ninan, K. N. see Reghunadhan Nair, C.P.: Vol. 155, pp. 1-99.
Noid, D. W. see Otaigbe, J. U.: Vol. 154, pp. 1-86.
Noid, D. W. see Sumpter, B. G.: Vol. 116, pp. 27-72.
Novac, B. see Grubbs, R.: Vol. 102, pp. 47-72.
Novikov, V. V. see Privalko, V. P.: Vol. 119, pp. 31-78.

O'Brien, D. F., Armitage, B. A., Bennett, D. E. and *Lamparski, H. G.*: Polymerization and Domain Formation in Lipid Assemblies. Vol. 126, pp. 53-84.
Ogasawara, M.: Application of Pulse Radiolysis to the Study of Polymers and Polymerizations. Vol.105, pp. 37-80.
Okabe, H. see Matsushige, K.: Vol. 125, pp. 147-186.
Okada, M.: Ring-Opening Polymerization of Bicyclic and Spiro Compounds. Reactivities and Polymerization Mechanisms. Vol. 102, pp. 1-46.
Okano, T.: Molecular Design of Temperature-Responsive Polymers as Intelligent Materials. Vol. 110, pp. 179-198.
Okay, O. see Funke, W.: Vol. 136, pp. 137-232.
Onuki, A.: Theory of Phase Transition in Polymer Gels. Vol. 109, pp. 63-120.

Oppermann W. see Holm, C.: Vol. 166, pp. 1-27.
Oppermann W. see Volk, N.: Vol. 166, pp. 29-65.
Osad'ko, I. S.: Selective Spectroscopy of Chromophore Doped Polymers and Glasses. Vol. 114, pp. 123-186.
Otaigbe, J. U., Barnes, M. D., Fukui, K., Sumpter, B. G., Noid, D. W.: Generation, Characterization, and Modeling of Polymer Micro- and Nano-Particles. Vol. 154, pp. 1-86.
Otsu, T., Matsumoto, A.: Controlled Synthesis of Polymers Using the Iniferter Technique: Developments in Living Radical Polymerization. Vol. 136, pp. 75-138.

de Pablo, J. J. see Leontidis, E.: Vol. 116, pp. 283-318.
Padias, A. B. see Penelle, J.: Vol. 102, pp. 73-104.
Pascault, J.-P. see Williams, R. J. J.: Vol. 128, pp. 95-156.
Pasch, H.: Analysis of Complex Polymers by Interaction Chromatography. Vol. 128, pp. 1-46.
Pasch, H.: Hyphenated Techniques in Liquid Chromatography of Polymers. Vol. 150, pp. 1-66.
Paul, W. see Baschnagel, J.: Vol. 152, p. 41-156.
Penczek, P. see Batog, A. E.: Vol. 144, pp. 49-114.
Penczek, P. see Bogdal, D.: Vol. 163, pp. 193-263.
Penelle, J., Hall, H. K., Padias, A. B. and Tanaka, H.: Captodative Olefins in Polymer Chemistry. Vol. 102, pp. 73-104.
Peppas, N. A. see Bell, C. L.: Vol. 122, pp. 125-176.
Peppas, N. A. see Hassan, C. M.: Vol. 153, pp. 37-65
Peppas, N. A. see Narasimhan, B.: Vol. 128, pp. 157-208.
Pet'ko, I. P. see Batog, A. E.: Vol. 144, pp. 49-114.
Pheyghambarian, N. see Kippelen, B.: Vol. 161, pp. 87-156.
Pichot, C. see Hunkeler, D.: Vol. 112, pp. 115-134.
Pielichowski, J. see Bogdal, D.: Vol. 163, pp. 193-263.
Pieper, T. see Kilian, H. G.: Vol. 108, pp. 49-90.
Pispas, S. see Pitsikalis, M.: Vol. 135, pp. 1-138.
Pispas, S. see Hadjichristidis: Vol. 142, pp. 71-128.
Pitsikalis, M., Pispas, S., Mays, J. W., Hadjichristidis, N.: Nonlinear Block Copolymer Architectures. Vol. 135, pp. 1-138.
Pitsikalis, M. see Hadjichristidis: Vol. 142, pp. 71-128.
Pleul, D. see Spange, S.: Vol. 165, pp. 43-78.
Pötschke, D. see Dingenouts, N.: Vol 144, pp. 1-48.
Pokrovskii, V. N.: The Mesoscopic Theory of the Slow Relaxation of Linear Macromolecules. Vol. 154, pp. 143-219.
Pospíšil, J.: Functionalized Oligomers and Polymers as Stabilizers for Conventional Polymers. Vol. 101, pp. 65-168.
Pospíšil, J.: Aromatic and Heterocyclic Amines in Polymer Stabilization. Vol. 124, pp. 87-190.
Powers, A. C. see Prokop, A.: Vol. 136, pp. 53-74.
Prasad, P. N. see Lin, T.-C.: Vol. 161, pp. 157-193.
Priddy, D. B.: Recent Advances in Styrene Polymerization.Vol. 111, pp. 67-114.
Priddy, D. B.: Thermal Discoloration Chemistry of Styrene-co-Acrylonitrile. Vol. 121, pp. 123-154.
Privalko, V. P. and *Novikov, V. V.:* Model Treatments of the Heat Conductivity of Heterogeneous Polymers.Vol. 119, pp 31-78.
Prociak, A see Bogdal, D.: Vol. 163, pp. 193-263
Prokop, A., Hunkeler, D., Powers, A. C., Whitesell, R. R., Wang, T. G.: Water Soluble Polymers for Immunoisolation II: Evaluation of Multicomponent Microencapsulation Systems. Vol. 136, pp. 53-74.
Prokop, A., Hunkeler, D., DiMari, S., Haralson, M. A., Wang, T. G.: Water Soluble Polymers for Immunoisolation I: Complex Coacervation and Cytotoxicity. Vol. 136, pp. 1-52.
Prokop, A., Kozlov, E., Carlesso, G and Davidsen, J. M.: Hydrogel-Based Colloidal Polymeric System for Protein and Drug Delivery: Physical and Chemical Characterization, Permeability Control and Applications. Vol. 160, pp. 119-174.

Pruitt, L. A.: The Effects of Radiation on the Structural and Mechanical Properties of Medical Polymers. Vol. 162, pp. 65-95.
Pudavar, H. E. see Lin, T.-C.: Vol. 161, pp. 157-193.
Pukánszky, B. and *Fekete, E.*: Adhesion and Surface Modification. Vol. 139, pp. 109 -154.
Putnam, D. and *Kopecek, J.*: Polymer Conjugates with Anticancer Acitivity. Vol. 122, pp. 55-124.

Quirk, R. P. and *Yoo, T., Lee, Y., M., Kim, J.* and *Lee, B.*: Applications of 1,1-Diphenylethylene Chemistry in Anionic Synthesis of Polymers with Controlled Structures. Vol. 153, pp. 67-162.

Ramaraj, R. and *Kaneko, M.*: Metal Complex in Polymer Membrane as a Model for Photosynthetic Oxygen Evolving Center. Vol. 123, pp. 215-242.
Rangarajan, B. see Scranton, A. B.: Vol. 122, pp. 1-54.
Ranucci, E. see Söderqvist Lindblad, M.: Vol. 157, pp. 139-161.
Raphaël, E. see Léger, L.: Vol. 138, pp. 185-226.
Reddinger, J. L. and *Reynolds, J. R.*: Molecular Engineering of p-Conjugated Polymers. Vol. 145, pp. 57-122.
Reghunadhan Nair, C. P., Mathew, D. and *Ninan, K. N.*, : Cyanate Ester Resins, Recent Developments. Vol. 155, pp. 1-99.
Reichert, K. H. see Hunkeler, D.: Vol. 112, pp. 115-134.
Rehahn, M., Mattice, W. L., Suter, U. W.: Rotational Isomeric State Models in Macromolecular Systems. Vol. 131/132, pp. 1-475.
Rehahn, M. see Bohrisch, J.: Vol. 165, pp. 1-41.
Rehahn, M. see Holm, C.: Vol. 166, pp. 1-27.
Reineker, P. see Holm, C.: Vol. 166, pp. 67-111.
Reynolds, J. R. see Reddinger, J. L.: Vol. 145, pp. 57-122.
Richter, D. see Ewen, B.: Vol. 134, pp.1-130.
Risse, W. see Grubbs, R.: Vol. 102, pp. 47-72.
Rivas, B. L. and *Geckeler, K. E.*: Synthesis and Metal Complexation of Poly(ethyleneimine) and Derivatives.Vol. 102, pp. 171-188.
Robin, J. J. see Boutevin, B.: Vol. 102, pp. 105-132.
Roe, R.-J.: MD Simulation Study of Glass Transition and Short Time Dynamics in Polymer Liquids. Vol. 116, pp. 111-114.
Roovers, J., Comanita, B.: Dendrimers and Dendrimer-Polymer Hybrids. Vol. 142, pp 179-228.
Rothon, R. N.: Mineral Fillers in Thermoplastics: Filler Manufacture and Characterisation.Vol. 139, pp. 67-108.
Rozenberg, B. A. see Williams, R. J. J.: Vol. 128, pp. 95-156.
Rühe, J., Ballauff, M., Biesalski, M., Dziezok, P., Gröhn, F., Johannsmann, D., Houbenov, N., Hugenberg, N., Konradi, R., Minko, S., Motornov, M., Netz, R. R., Schmidt, M., Seidel, C., Stamm, M., Stephan, T., Usov, D. and *Zhang, H.*: Polyelectrolyte Brushes. Vol. 165, pp. 79-150.
Ruckenstein, E.: Concentrated Emulsion Polymerization. Vol. 127, pp. 1-58.
Ruiz-Taylor, L. see Mathieu, H. J.: Vol. 162, pp. 1-35.
Rusanov, A. L.: Novel Bis (Naphtalic Anhydrides) and Their Polyheteroarylenes with Improved Processability. Vol. 111, pp. 115-176.
Russel, T. P. see Hedrick, J. L.: Vol. 141, pp. 1-44.
Rychlý, J. see Lazár, M.: Vol. 102, pp. 189-222.
Ryner, M. see Stridsberg, K. M.: Vol. 157, pp. 2751.
Ryzhov, V. A. see Bershtein, V. A.: Vol. 114, pp. 43-122.

Sabsai, O. Y. see Barshtein, G. R.: Vol. 101, pp. 1-28.
Saburov, V. V. see Zubov, V. P.: Vol. 104, pp. 135-176.
Saito, S., Konno, M. and *Inomata, H.*: Volume Phase Transition of N-Alkylacrylamide Gels. Vol. 109, pp. 207-232.
Samsonov, G. V. and *Kuznetsova, N. P.*: Crosslinked Polyelectrolytes in Biology. Vol. 104, pp. 1-50.
Santa Cruz, C. see Baltá-Calleja, F. J.: Vol. 108, pp. 1-48.
Santos, S. see Baschnagel, J.: Vol. 152, p. 41-156.

Sato, T. and *Teramoto, A.*: Concentrated Solutions of Liquid-Christalline Polymers. Vol. 126, pp. 85-162.
Schaller, C. see Bohrisch, J.: Vol. 165, pp. 1-41.
Schäfer R. see Köhler, W.: Vol. 151, pp. 1-59.
Scherf, U. and *Müllen, K.*: The Synthesis of Ladder Polymers.Vol. 123, pp. 1-40.
Schmidt, M. see Förster, S.: Vol. 120, pp. 51-134.
Schmidt, M. see Rühe, J.: Vol. 165, pp. 79-150.
Schmidt, M. see Volk, N.: Vol. 166, pp. 29-65.
Scholz, M.: Effects of Ion Radiation on Cells and Tissues. Vol. 162, pp. 97-158.
Schopf, G. and *Koßmehl, G.*: Polythiophenes - Electrically Conductive Polymers. Vol. 129, pp. 1-145.
Schulz, E. see Munz, M.: Vol. 164, pp. 97-210.
Sturm, H. see Munz, M.: Vol. 164, pp. 87-210.
Schweizer, K. S.: Prism Theory of the Structure, Thermodynamics, and Phase Transitions of Polymer Liquids and Alloys. Vol. 116, pp. 319-378.
Scranton, A. B., Rangarajan, B. and *Klier, J.*: Biomedical Applications of Polyelectrolytes. Vol. 122, pp. 1-54.
Sefton, M. V. and *Stevenson, W. T. K.*: Microencapsulation of Live Animal Cells Using Polycrylates. Vol.107, pp. 143-198.
Seidel, C. see Holm, C.: Vol. 166, pp. 67-111.
Seidel, C. see Rühe, J.: Vol. 165, pp. 79-150.
Shamanin, V. V.: Bases of the Axiomatic Theory of Addition Polymerization. Vol. 112, pp. 135-180.
Sheiko, S. S.: Imaging of Polymers Using Scanning Force Microscopy: From Superstructures to Individual Molecules. Vol. 151, pp. 61-174.
Sherrington, D. C. see Cameron, N. R.,Vol. 126, pp. 163-214.
Sherrington, D. C. see Lin, J.: Vol. 111, pp. 177-220.
Sherrington, D. C. see Steinke, J.: Vol. 123, pp. 81-126.
Shibayama, M. see Tanaka, T.: Vol. 109, pp. 1-62.
Shiga, T.: Deformation and Viscoelastic Behavior of Polymer Gels in Electric Fields. Vol. 134, pp. 131-164.
Shim, H.-K., Jin, J.: Light-Emitting Characteristics of Conjugated Polymers. Vol. 158, pp. 191-241.
Shoda, S. see Kobayashi, S.: Vol. 121, pp. 1-30.
Siegel, R. A.: Hydrophobic Weak Polyelectrolyte Gels: Studies of Swelling Equilibria and Kinetics. Vol. 109, pp. 233-268.
Silvestre, F. see Calmon-Decriaud, A.: Vol. 207, pp. 207-226.
Sillion, B. see Mison, P.: Vol. 140, pp. 137-180.
Simon, F. see Spange, S.: Vol. 165, pp. 43-78.
Singh, R. P. see Sivaram, S.: Vol. 101, pp. 169-216.
Sinha Ray, S. see Biswas, M: Vol. 155, pp. 167-221.
Sivaram, S. and *Singh, R. P.*: Degradation and Stabilization of Ethylene-Propylene Copolymers and Their Blends: A Critical Review. Vol. 101, pp. 169-216.
Söderqvist Lindblad, M., Liu, Y., Albertsson, A.-C., Ranucci, E., Karlsson, S.: Polymer from Renewable Resources.Vol. 157, pp. 139-161
Spange, S., Meyer, T., Voigt, I., Eschner, M., Estel, K., Pleul, D. and *Simon, F.*: Poly(Vinylformamide-co-Vinylamine)/Inorganic Oxid Hybrid Materials. Vol. 165, pp. 43-78.
Stamm, M. see Möhwald, H.: Vol. 165, pp. 151-175.
Stamm, M. see Rühe, J.: Vol. 165, pp. 79-150.
Starodybtzev, S. see Khokhlov, A.: Vol. 109, pp. 121-172.
Stegeman, G. I. see Canva, M.: Vol. 158, pp. 87-121.
Steinke, J., Sherrington, D. C. and *Dunkin, I. R.*: Imprinting of Synthetic Polymers Using Molecular Templates. Vol. 123, pp. 81-126.
Stenzenberger, H. D.: Addition Polyimides. Vol. 117, pp. 165-220.
Stephan, T. see Rühe, J.: Vol. 165, pp. 79-150.
Stevenson,W. T. K. see Sefton, M. V.: Vol. 107, pp. 143-198.

Stridsberg, K. M., Ryner, M., Albertsson, A.-C.: Controlled Ring-Opening Polymerization: Polymers with Designed Macromoleculars Architecture. Vol. 157, pp. 2751.
Sturm, H. see *Munz, M.:* Vol. 164, pp. 87–210.
Suematsu, K.: Recent Progress of Gel Theory: Ring, Excluded Volume, and Dimension. Vol. 156, pp. 136-214.
Sumpter, B. G., Noid, D. W., Liang, G. L. and *Wunderlich, B.:* Atomistic Dynamics of Macromolecular Crystals. Vol. 116, pp. 27-72.
Sumpter, B. G. see *Otaigbe, J.U.:* Vol. 154, pp. 1-86.
Sugimoto, H. and *Inoue, S.:* Polymerization by Metalloporphyrin and Related Complexes. Vol. 146, pp. 39-120.
Suter, U. W. see *Gusev, A. A.:* Vol. 116, pp. 207-248.
Suter, U. W. see *Leontidis, E.:* Vol. 116, pp. 283-318.
Suter, U. W. see *Rehahn, M.:* Vol. 131/132, pp. 1-475.
Suter, U. W. see *Baschnagel, J.:* Vol. 152, p. 41-156.
Suzuki, A.: Phase Transition in Gels of Sub-Millimeter Size Induced by Interaction with Stimuli. Vol. 110, pp. 199-240.
Suzuki, A. and *Hirasa, O.:* An Approach to Artifical Muscle by Polymer Gels due to Micro-Phase Separation. Vol. 110, pp. 241-262.
Swiatkiewicz, J. see *Lin, T.-C.:* Vol. 161, pp. 157-193.

Tagawa, S.: Radiation Effects on Ion Beams on Polymers.Vol. 105, pp. 99-116.
Tan, K. L. see *Kang, E. T.:* Vol. 106, pp. 135-190.
Tanaka, H. and *Shibayama, M.:* Phase Transition and Related Phenomena of Polymer Gels.Vol. 109, pp. 1-62.
Tanaka, T. see *Penelle, J.:* Vol. 102, pp. 73-104.
Tauer, K. see *Guyot, A.:* Vol. 111, pp. 43-66.
Teramoto, A. see *Sato, T.:* Vol. 126, pp. 85-162.
Terent'eva, J. P. and *Fridman, M. L.:* Compositions Based on Aminoresins. Vol. 101, pp. 29-64.
Theodorou, D. N. see *Dodd, L. R.:* Vol. 116, pp. 249-282.
Thomson, R. C., Wake, M. C., Yaszemski, M. J. and *Mikos, A. G.:* Biodegradable Polymer Scaffolds to Regenerate Organs. Vol. 122, pp. 245-274.
Thünemann, A. F., Müller, M., Dautzenberg, H., Joanny, J.-F., Löwen, H.: Polyelectrolyte complexes. Vol. 166, pp. 113-171.
Tieke, B. see v. *Klitzing, R.:* Vol. 165, pp. 177-210.
Tokita, M.: Friction Between Polymer Networks of Gels and Solvent. Vol. 110, pp. 27-48.
Traser, S. see *Bohrisch, J.:* Vol. 165, pp. 1-41.
Tries, V. see *Baschnagel, J.:* Vol. 152, p. 41-156.
Tsuruta, T.: Contemporary Topics in Polymeric Materials for Biomedical Applications. Vol. 126, pp. 1-52.

Usov, D. see *Rühe, J.:* Vol. 165, pp. 79-150.
Uyama, H. see *Kobayashi, S.:* Vol. 121, pp. 1-30.
Uyama, Y: Surface Modification of Polymers by Grafting. Vol. 137, pp. 1-40.

Varma, I. K. see *Albertsson, A.-C.:* Vol. 157, pp. 99-138.
Vasilevskaya, V. see *Khokhlov, A.:* Vol. 109, pp. 121-172.
Vaskova, V. see *Hunkeler, D.:* Vol.: 112, pp. 115-134.
Verdugo, P.: Polymer Gel Phase Transition in Condensation-Decondensation of Secretory Products. Vol. 110, pp. 145-156.
Vettegren, V. I. see *Bronnikov, S. V.:* Vol. 125, pp. 103-146.
Vilgis, T. A. see *Holm, C.:* Vol. 166, pp. 67-111.
Viovy, J.-L. and *Lesec, J.:* Separation of Macromolecules in Gels: Permeation Chromatography and Electrophoresis. Vol. 114, pp. 1-42.
Vlahos, C. see *Hadjichristidis, N.:* Vol. 142, pp. 71-128.
Voigt, I. see *Spange, S.:* Vol. 165, pp. 43-78.
Volk, N., Vollmer, D., Schmidt, M., Oppermann, W., Huber, K.: Conformation and Phase Diagrams of Flexible Polyelectrolytes. Vol. 166, pp. 29-65.

Volksen, W.: Condensation Polyimides: Synthesis, Solution Behavior, and Imidization Characteristics. Vol. 117, pp. 111-164.
Volksen, W. see Hedrick, J. L.: Vol. 141, pp. 1-44.
Volksen, W. see Hedrick, J. L.: Vol. 147, pp. 61-112.
Vollmer, D. see Volk N.: Vol. 166, pp. 29-65.

Wake, M. C. see Thomson, R. C.: Vol. 122, pp. 245-274.
Wandrey C., Hernández-Barajas, J. and *Hunkeler, D.*: Diallyldimethylammonium Chloride and its Polymers. Vol. 145, pp. 123-182.
Wang, K. L. see Cussler, E. L.: Vol. 110, pp. 67-80.
Wang, S.-Q.: Molecular Transitions and Dynamics at Polymer/Wall Interfaces: Origins of Flow Instabilities and Wall Slip. Vol. 138, pp. 227-276.
Wang, S.-Q. see Bhargava, R.: Vol. 163, pp. 137-191.
Wang, T. G. see Prokop, A.: Vol. 136, pp.1-52; 53-74.
Wang, X. see Lin, T.-C.: Vol. 161, pp. 157-193.
Whitesell, R. R. see Prokop, A.: Vol. 136, pp. 53-74.
Williams, R. J. J., Rozenberg, B. A., Pascault, J.-P.: Reaction Induced Phase Separation in Modified Thermosetting Polymers. Vol. 128, pp. 95-156.
Winkler, R. G. see Holm, C.: Vol. 166, pp. 67-111.
Winter, H. H., Mours, M.: Rheology of Polymers Near Liquid-Solid Transitions. Vol. 134, pp. 165-234.
Wittmeyer, P. see Bohrisch, J.: Vol. 165, pp. 1-41.
Wu, C.: Laser Light Scattering Characterization of Special Intractable Macromolecules in Solution. Vol 137, pp. 103-134.
Wunderlich, B. see Sumpter, B. G.: Vol. 116, pp. 27-72.

Xiang, M. see Jiang, M.: Vol. 146, pp. 121-194.
Xie, T. Y. see Hunkeler, D.: Vol. 112, pp. 115-134.
Xu, Z., Hadjichristidis, N., Fetters, L. J. and *Mays, J. W.*: Structure/Chain-Flexibility Relationships of Polymers. Vol. 120, pp. 1-50.

Yagci, Y. and *Endo, T.*: N-Benzyl and N-Alkoxy Pyridium Salts as Thermal and Photochemical Initiators for Cationic Polymerization. Vol. 127, pp. 59-86.
Yannas, I. V.: Tissue Regeneration Templates Based on Collagen-Glycosaminoglycan Copolymers. Vol. 122, pp. 219-244.
Yang, J. S. see Jo, W. H.: Vol. 156, pp. 1-52.
Yamaoka, H.: Polymer Materials for Fusion Reactors. Vol. 105, pp. 117-144.
Yasuda, H. and *Ihara, E.*: Rare Earth Metal-Initiated Living Polymerizations of Polar and Nonpolar Monomers. Vol. 133, pp. 53-102.
Yaszemski, M. J. see Thomson, R. C.: Vol. 122, pp. 245-274.
Yoo, T. see Quirk, R. P.: Vol. 153, pp. 67-162.
Yoon, D. Y. see Hedrick, J. L.: Vol. 141, pp. 1-44.
Yoshida, H. and *Ichikawa, T.*: Electron Spin Studies of Free Radicals in Irradiated Polymers. Vol. 105, pp. 3-36.

Zhang, H. see Rühe, J.: Vol. 165, pp. 79-150.
Zhou, H. see Jiang, M.: Vol. 146, pp. 121-194.
Zubov, V . P., Ivanov, A. E. and *Saburov, V. V.* : Polymer-Coated Adsorbents for the Separation of Biopolymers and Particles. Vol. 104, pp. 135-176.

Subject Index

ABC triblock copolymer   192, 196–197, 206
Acrylamide   118
Adsorption   126
AFM   93, 151, 159, 180
Aggregation number   120-121, 165, 176–177, 195
Alkaline earth cation   32
Alkyl sulfate   138
Amphiphile   135, 192, 201-204
Amphiphilic graft copolymer   203–204
Arginine   142
ASAXS   21
Atomic force microscopy (AFM)   93, 151, 159, 180
ATR-FTIR   131
–, in-situ   126

Bead-spring chain   98, 102
Birefringence, electric   1, 10
Bjerrum length   75, 84, 162
Block copolyampholyte   191, 193
– –, amphiphilic   192, 201
Block copolymer   136, 154, 192, 196–206
– – micelles   159, 201, 206
– –, microphase-separated   154
Block polyampholyte   161, 165
Blood plasma protein   134
Boltzmann factor   5
Bragg reflection   153
Bragg spacing   141, 148
Brush   179, 201
–, core-shell   202
–, – cylinder   201
–, osmotic   179
–, salted   179
4-Butylphenylmaleamic acid   137

CD spectroscopy   39, 125
Cell model   25
Cellulose   126
Charge chemical potential   94
Charge density   118

Charge fraction   165
Charge neutralization   118
Chloride counterion   23
Circular dichroism (CD)   139, 149
Coil dimension   32, 42, 60
Coil-to-globule transition   123
Colloidal salt stability   122
Colloidal titration   118
Colloids   115
–, polymeric   196
Complex formation   116, 118, 122
Concentration fluctuation   162
Conductivity   13, 15, 46–48, 118
Conductometry   118
Conformational adaptation   117
Cooperative binding   115, 135
Copolyampholyte, triblock   192–194
Core-shell brush   202
– cylinder   201
Correlation functions   71–75
Correlation length   92, 153
Coulomb interaction   78, 166
Counterion condensation   4, 32, 39, 43, 74, 95, 116, 166
–, multivalent   8
Counterion-counterion potential   75
Cryogenic transmission electron microscopy   199
Cylinder micelle, unimolecular   197
Cylindrical-shaped particle   137

Debye-Hückel   71, 76-77, 105, 161
– hole-cavity (DHHC)   8, 18-19, 71
– screening length   104, 162, 182
Demixing transition   163
Density approximation, local   70, 71
Density functional extensions   70
DHHC   18–19, 71
N,N'-Diallyl-N''-dimethylammonium chloride   137
Diblock copolymer   161, 165, 177, 202
Dielectric boundary   105

Differential scanning calorimetry (DSC) 138
Dilute solutions 73
Diluted system 120
Discotic columnar structure 135
Disk-shaped particle 137
Dispersion 113
Disproportionation 117
DNA 3, 125
Dodecanoic acid 136, 137
Donnan equilibrium 181
Double-layer stack 155
Doughnut-shape 159
Doxorubicin 144
Drug delivery/targeting system 137
Dynamic light scattering (DLS) 119, 179, 194, 196, 199, 204

Effective potential 75
Einstein-Stokes equation 119
Electric birefringence 1, 10
Electron density, radial excess 21
Ellipticity, molar 150

Fatty acid 137
Filaments 184
Flocculation 118, 120, 122, 128
Flory interaction parameter 162, 166
Flow field-flow fractionation, asymmetric 199
Fluorescence spectroscopy 137
Fluorometry 204
Fluorosurfactant complex 148
Force extension relation 93
Form factor 100
Free energy 161
Freeze-drying 141
Frozen structure 120
Frustration 142

Gaussian chain 97
– polyelectrolyte 95
– polymer 162
Glass-transition temperature 138
Globular conformation 87
Good solvent 95
Graft copolymer 177, 197, 203–205, 207
Grafting distance 176

Helix-coil transition 140, 147, 149–151
Histidine 140, 142
Human serum albumin 133
Hybrid 202
Hydratation shell 20

Hydrodynamic radius/diameter 119, 136, 194, 200
Hydrophobic effect 86

Imidazole group 140
Immobilization 124, 138
Integral equation 8
Interaction, non-electrostatic 162
–, specific 29, 31
Interpolymer complex 116
Iodine 23
Ion distribution function, integrated 17
Ionic strength 118, 121, 165

Janus micelle 197, 199, 207

Kerr's law 10
Kiessig fringes 155

Ladder-egg structure 113
Ladder-like structure 117
Lamellar particle structure 142
Lamellar structure/thickness 135, 141, 143
Langevin thermostat 102
Latex 128
Layer-by-layer technique 115, 128
Light scattering 203
Limiting law, Manning 7
Lipid 135
Liquid-like order 73
Local density approximation 71
Lorentzian peak profile 141
Low-dielectric substrate 107
Lysine 142

Macroion 116
Maleamic acid 137
Manning condensation 70
Manning parameter 7, 17
Mesomorphous structure 135
Mesophase 113, 135
Methacrylic acid 136
Methyl methacrylate 136
Micellar networks, formation 185
Micelles 144, 175–176, 191–196, 201–203
–, core-shell 144, 148, 152
–, cylinder 202
–, cylindrical 12
–, density profile 182
–, dressed 137
–, frozen 175
–, supermicelle 197–200, 207
–, triblock cololymer 191
–, unimolecular 197
–, wormlike 201

Micellization 165
Microgel 123
Microphase-separated block copolymer 154
Mixing ratio, molar 116, 118
Molar ellipticity, mean 150
Molar mixing ratio 116, 118
Molecular dynamics (MD) 4
Monte Carlo (MC) 98
Multilayer 113, 128, 133, 156

Nanodispersion 141
Nanoparticle, fluorinated 136
–/nanostructure 113, 115, 136, 141–142, 161, 196–197
Nanostructure, lamellar 141
Na-polystyrene sulfonate 20
Network 93, 184, 189
NMR, ^{1}H-NMR relaxation 133

Oligopeptide 135
One component plasma (OCP) 71
One-loop approximation 163
Ornstein-Zernike equation 72
Osmotic brush 179
– coefficient 4, 16, 17, 43–45, 49
– pressure 7, 162
Overcharging 116, 122

P3 M 102
Particle formation/structure 136, 143
PB, linearized 8
PB-cell model 22
PB-theory 18–19
Pearl-necklace 97
PEC formation 113, 115, 120
PEE-PSSH micelles, density profiles 184
PEI structure 136, 152, 161
PEI-retinoate complex 152
PEO 136, 145
– corona 151
PEO-PLL 138
Perfluorododecanoic acid 136
Perturbation approach 78
PE-surf 113, 135
– nanoparticle 136
pH 98
Phase separation 71, 183
Phenylmaleamic acid 137
Photocrosslinking 131
Photoluminescence 135
PLA 141
Plasma, one-component 71
PLH 141
PLL 141

PMMA 192
Poisson-Boltzmann (PB) theory 1, 5, 19, 67, 70
Polarizability, anisotropy 13
Poloxamer 188 141, 158
Poly(acrylic acid) 125–126
Poly(allylamine) 133
Poly(L-arginine) 138
– retinoate 140
Poly(diallyldimethylammonium chloride) 118, 126, 138
Poly(ethylene imine) (PEI) 136–137, 152
Poly(ethylene oxide) (PEO) 136, 145
Poly(ethylene oxide)-b-poly(L-lysine) 144, 152
Poly(ethyleneimine) 126
Poly(L-histidine) 138
– retinoate 140
Poly(ionene-6,3) 138
Poly(N-isopropylacrylamide) 123
Poly(L-lysine) 125, 131, 138, 145
– retinoate 140
Poly(maleic acid-co-ethylene) 125
Poly(maleic acid-co-methylstyrene) 126, 131
Poly(N-methyl-4-vinyl-pyridinium) 138
Poly(styrene) 136
Poly(styrene sulfonate) 20, 118, 136
Poly(vinyl sulfate) 125, 131, 134
Poly(4-vinylbenzyl-trimethylammonium chloride) 118
Polyacid 116
Polyampholyte 136
Polyanion/polycation 115
Polybase 116
Polydispersity 141–142
Polyelectrolyte 69, 115, 135
–, annealed 94–98
– block copolymer 175
–, cationic 1
– complex 113, 122, 161, 193
– corona 165
–, flexible 76
–, Gaussian 95
– layer 131
– multilayer 128, 161, 164
–, rigid-rod 131
–, rod-like 1, 3, 18, 25
– star 166
–, strong 94
–, strongly charged, ion distribution 101
– vesicles 189
–, weak 94
Polyelectrolyte–surfactant complex 113, 135, 166

Polyethyleneimine (PEI) 136, 152
Polyions, equilibrium properties, MC 98
Polypropylene 133
Polystyrene sulfonate 20, 118, 136
Porod's law 154
Potential, effective 75
–, medium-induced 77
Potentiometry 118
PPP 3, 4, 9, 18
Precipitation 162
Precursor polymer 10
Primitive model 72
– –, restricted 4, 7
PRISM (polymer-reference-interaction-site model) 72
Protein 124, 133
– resistance 134
Proton transfer 149
PS-P4VPMeI 176
PS-PAAc 176
PtBS-PSSNa 176
Pyrene 137

Q_{10} 137

Radial distribution function 70
Radial excess electron density 21
Radius of gyration 119
Random-phase approximation 129
Rayleigh instability 87
Resonant contribution 22
Retinoic acid 113, 137, 145, 151, 156
Rigid rod 99, 131
Rod-like molecule 9
RPM cell model 16

Salted brush 179
Salting in/out 29, 31, 37, 46, 49–50, 54
SAXS measurement 1, 18, 21
Scaling 67, 82
Scanning electron microscopy (SEM) 199
Scanning force microscopy (SFM) 199, 202
Scattering function 92
Scattering intensity I(q) 21
Schulz-Flory distribution 11
Scrambled-egg model 113, 117
Secondary aggregation 120–123
Self-assembly 115
Semiflexible chain 77
β-Sheet 147, 150
Silica surface 126
Simulation 98
Single chain form factor 91

Small-angle neutron scattering (SANS) 176, 179, 204
Small-angle X-ray scattering (SAXS) 4, 137, 143, 146, 151, 153, 179
Solid-state complex 138
Solid-state structure 115
Solvent 22
–, poor 97, 104
Soy lecithin 138
Spheres, soft 187
Spin-coat technique 155
Star, polyelectrolyte 166
Static structure factor 82
Statistic copolymer 136
Steric interaction 188
Stop flow measurement 117
Structure function 84
Structures-*within*-structures 135
Styrylmethyl(trimethyl)ammonium chloride 136
Substitution reaction 117
Supermicelle 197–200, 207
Superstructure 207
Supramolecular order 158
Surface-compartmentalized nanoparticle 197
Surfactant 138
Surfactant, fluorinated 155
Suzuki coupling 9
Swelling 120

Tart-type particle structure 143
TEM, cryogenic 183, 199
–, polyelectrolyte micelles 177, 179, 184, 196, 203
Tensiometry 204
Thermodynamic equilibrium 117
Titration, potentiometric 194
Topology, nonlinear 196
Toroid structure 159
Transfection 125
Translational entropy 162, 166
Transmission electron microscopy (TEM) 177, 179, 184, 196, 203
Triblock copolyampholyte 192–194
Triblock copolymer 141, 177
Triiodothyronine 137
Turbidity 118

Ultracentrifuge 20
Ultra-thin film 161
UV spectroscopy 118

Variational theory 166
Vesicles 189, 191, 194

Subject Index

Virus, rod-like 25
Viscometry 119

Water immobilization 133
Water-soluble complex 138
Wide-angle X-ray scattering (WAXS) 140
Wormlike micelle 201

X-ray scattering/reflectivity 1

Zeta potential 136–137
Zipper mechanism, cooperative 115, 117

Printing: Saladruck, Berlin
Binding: Stein+Lehmann, Berlin